AN INTRODUCTION TO ASTRONOMY

Brian Daugherty

Websites
http://www.maccer.co.uk/main   umbrella site for all my Astronomy and Maths sites

Other books by Brian Daugherty

The Mensa Quiz Book, Carlton Books, 978-1787390218

GCSE Mathematics, An Informal Overview,   978-0-244-68040-4

Foreword

There is no need to rush out and buy a telescope in order to initiate an interest in astronomy. If you already have a set of binoculars then they will suffice for beginners, even if they are just 'general-purpose' and maybe unsuitable for more advanced observing. There are indeed a few advantages of binoculars for beginners

- Two eyes are better than one originally
- Images in binoculars are the right way up
- Most telescopes have magnifications > 25 and will therefore have to be mounted rigidly to see the detail at this magnification. The field of view at such magnifications will be small.

I need to give the usual warning about not viewing the Sun using an optical instrument and hope that once will suffice for anyone with any sense. Infra-red radiation will cause immediate damage to your eyes if viewed through binoculars or a telescope.

In many cases you will need to wrap up well because cold conditions usually go hand in hand with good astronomical seeing, and obviously winter provides more hours of viewing than the Summer.[1] But as stated by Pam Spence[2]: 'You don't <u>have</u> to freeze to death at three o'clock in the morning just to see the fuzzy outline of Comet *XXX*, but if you do, it is a marvelous feeling'.

The images you will see will not be as colorful as some of the pictures you often see in books. These pictures need not be artificial colors as such but could just be long-exposure photographs which are capable of showing colors that the human eye is not capable of seeing. For example, there are modern-day camera films capable of taking a picture of grass at night and producing a definite green color – we do know that this is a natural color, grass really is green, but no matter how hard we look at it during the night our eyes will never detect this green color.

It is not too unusual for beginners to see nothing at all in a telescope when more experienced observers have just told them that such-and-such an object

---

[1] And during the mid Summer, the 'dark' night sky is actually permanently in twilight – it never gets as dark as at other times of the year
[2] Editor of 'Astronomy Now' at the time

is in the field of view. This will mostly be because their eyes have not become dark-adapted enough. Dark adaption is something you experience in real-life – on entering a darkened room, you might not be able to initially see anything but given time your seeing will improve, in all probability. So before using a telescope you do need to give your eyes time to adjust. If you need any light at all it should be red light[3] and should not be left on for longer than you need it. According to circumstances, you might want to adopt other measures, e.g. wearing sunglasses when not observing. One person I know recommends wearing an eye-patch, like a pirate, for about 20 minutes or so before starting observations. Quite often you will see advice to cover your head with a dark cloth to block out any stray light

Nevertheless it is true that people vary – some are more sensitive to the detection of faint objects than others. Furthermore the same person can have varying sensibilities between their left and right eyes.

You need to be aware that 'averted vision' can sometimes result in seeing an object better than looking at it directly. A good example of this is the Pleiades Cluster in Taurus, a very noticeable object to those in the know but possibly less noticeable to others for the reason just given.

Counteracting light pollution is a subject dear to astronomers' hearts but it is actually easier for beginners to learn the subject in a polluted sky where the brighter stars are more obvious. I could well imagine that it might be very difficult for beginners in very clear, unpolluted skies.

To learn the sky properly you will need to observe regularly over at least a year – even those constellations that are in the sky all year long will adopt different orientations over the space of a year

Metric units are used in the text, as has always been the case in scientific circles during my time – which is not surprising given that the system itself was presented as long ago as 1800 and prepared by a committee of world-class mathematicians such as Lagrange, Laplace, Carnot, Laguerre etc. The word 'billion' is used instead of thousand million as is also standard in scientific circles. This is consistent with the conventions established by the metric system whereby outside the 'basic' range, new names are given for each increase/decrease by a factor of 1 000[4] – the phrase 'thousand million' is basically too long to be used frequently

The material discussed here is typical of the topics that would be discussed in a magazine like *Astronomy Now* which is directed to amateurs. Be aware that the magazine *Sky and Telescope*, which is also common, is directed at a higher level.

Brian Daugherty

---

[3] If all else fails, stick some red plastic around a conventional white torch
[4] strictly speaking. in the metric system it is new prefixes that are added for each increase/decrease by a factor of 1000

Chapter 1 Introduction

Imagine the totally false but useful idea that the Earth is at the center of the Universe, surrounded by a Celestial Sphere upon which all the stars are fixed[5]. Because of the 'clockwork' nature of this model, we could even attempt to assign the Sun to a place on the Celestial Sphere despite not being able to see both Sun and stars simultaneously.

The large-scale motion of the heavens that we observe during the night[6] is then due to the rotation of this Celestial Sphere about us.

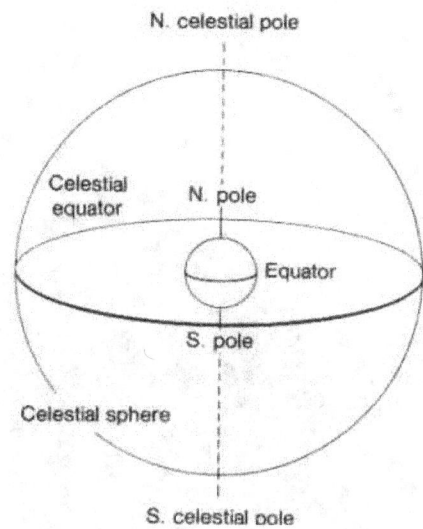

Figure 1 - the Celestial Sphere

However, over several nights, it will become apparent that several bodies are in fact moving *relative to the Celestial Sphere*. For starters, the Sun is moving, as is the Moon and the planets[7].

The Sun moves once round the Celestial Sphere (i.e. it circles it) in one year, meaning that within 24 hours it will move approximately 1 degree[8]. Additionally, as it moves around the Celestial Sphere, it will be lower in the sky in Winter and higher in the sky in Summer - it alters its **altitude**[9], The path that the Sun traces on the Celestial Sphere is called the **ecliptic**.

A recommended tool to aid observations of the sky is a **planisphere** (as shown in *Figure 2* - Planisphere below). The full instructions for use are normally on the back (as is the case for the Philip's example shown) but you can see the basic idea from the diagram - the view of the sky shown can be seen from the information on the outer edge to correspond to the view at different times on different dates. For the particular view in the diagram, we can see (to quote just three examples of many) that it corresponds to the view at 2400 sometime in December, 1800 sometime in March or 1200 sometime in June (although in the last case it obviously can't actually be seen visually).

---

[5] although this is false, our brain does seem actually perceive the heavens in this manner. This produces the 'Moon illusion' whereby the Moon looks larger when it is near the horizon than when it is high in the sky, despite the fact that it is actually smaller. By virtue of us subconsciously thinking of the Moon as being 'attached' to a dome, you can perhaps appreciate why we would therefore think it was bigger when low down and hence 'further away on the dome'.

[6] and for the Sun (and often the Moon), the motion we observe during the day

[7] Just to be clear here – the motion we see of the stars, Sun, Moon and planets across the sky in a single night or day is due to the 'clockwork' motion of this Celestial Sphere'. But the motion I am now referring to in this paragraph is motion on the Celestial Sphere itself. The stars stay where they are on the Celestial Sphere but the Sun and Moon and planets move around the Celestial Sphere.

[8] Because a circle has 360 degrees and a year has about the same number of days .

[9] altitude is defined as the angle between the horizon and the object

Figure 2 - Planisphere

Additionally, you can see an obvious arc within the viewer[10]. This represents the ecliptic. To actually assign a point on the ecliptic to a particular date, you need to consider a line from the central pin to a particular date, and where this line intersects the ecliptic is where the Sun will be on that particular date.

Since the center-pin represents the location of the North Star (or Polaris as astronomers tend to call it), you might be able to perceive that this view does indeed show the Sun at a higher altitude in June than at other relevant dates for this particular view[11].

Although in daytime you can't actually see the stars behind the Sun[12], we can nevertheless calculate what constellation it is in. Such is the clockwork nature of this model that this was something that even the Ancients could calculate –

---

[10] From near the word 'Eastern' to near the word 'Western'
[11] Because then the Sun will be closer to the Pole Star
[12] unless there is a solar eclipse

and this provided the origins for the dreaded subject of astrology. The star signs are intended to indicate that the Sun is in this particular constellation at the particular time in question. Since the time astrology was invented, the Earth's 'wobble' (or precession) has disrupted the original scheme, but the star signs can still be of some use for remembering the zodiacal sequence - they are only about one month out. In astrology, Taurus corresponds to April/May. In reality the Sun is in Taurus in May/June (and consequently it will be best placed for night-sky viewers about six months later, and since the Sun was fairly high when it was in Taurus this is an indication that Taurus will be fairly high in the night sky - the fact that the Sun is low down when it is in the likes of Sagittarius is an indication that Sagittarius will be low down in the night sky).

Nowadays the **zodiac** (the constellations through which the Sun passes during one complete year) actually comprises 13 constellations, and the time it spends in each is highly variable, far removed from the almost equal divisions of astrology.

When you think about the way the Solar System is arranged (i.e. actually arranged in the real model), then you can see that the planets (when they can be seen) will also be close to the ecliptic. And as chance would have it the Moon is to be seen near the ecliptic as well, although there is no actual reason why a natural satellite should orbit a planet in such a manner – it is just a coincidence[13].

Now the North Star is represented by the center-pin on the planisphere and can therefore seen to be stationary in the sky while the other stars appear to 'circle' around it during the period of a night[14]

Usually in a class I would now ask what two major advantages Polaris offered to mariners. The first is normally well known – unsurprisingly it points to the North. The second advantage is usually not so forthcoming but it is this : the altitude of Polaris above the horizon is equal to the latitude of the observer. You might have heard of the difficulties mariners had in determining their longitude, but latitude was never such of a problem

Coordinates

The Celestial Sphere has a coordinate system projected on it, which is essentially just a projection of the Earth's longitude and latitude. The 'latitude' component is called **declination** and is similar to the Earth's latitude insofar as Polaris lies at 90° and we have a **Celestial Equator** at 0°. One difference is that northern declinations are indicated by a plus sign (+54°, for example) while Southern declinations are indicated by a minus sign (-46°).

---

[13] It is also a coincidence that it is about the same size in the Sky as the Sun, thus being capable of producing total solar eclipses. The Moon is moving away from the Earth so these total eclipses will eventually come to an end.
[14] Again this is a simplified model. The North Star is not exactly at the point of 'zero rotation' and does actually itself 'circle' about the pole - just a little

Referring to Figure 1, anyone situated at say 50° N on the Earth will be able to see below the Celestial Equator when looking South but will not be able to see anywhere as far as the Celestial Equator when looking North.

Note also half of the ecliptic will be North of the Celestial Equator and the other half will be South of the Celestial Equator.

The analog of longitude is called **right ascension**. This is given in units of hours and minutes, derived from the time taken for the Earth to rotate[15]. So a full 'circle' corresponds to 24 hours of right ascension. A difference of one hour of right ascension will correspond to the distance that the Earth will rotate in one hour. Reverting to some simple mathematics for a moment, this will correspond in 'more conventional' mathematical units to 360/24 = 15°.

Just like longitude, right ascension requires an arbitrary 'starting point' to be designated as the zero. For right ascension, this 'zero' turns out not to be so arbitrary as the Greenwich Meridian[16]. The 'zero' for right ascension is called **the First Point of Aries** and is represented by a meridian that goes through the point of intersection of the ecliptic and the Celestial Equator corresponding to where the Sun is on, or around, March 21st. I need to stress that it is the intersection relating to the March event because of course the ecliptic intersects the celestial equator twice.

Right Ascension is measured off eastwards from the First Point of Aries.

The first thing you will notice when you refer to the planisphere is that the first Point of Aries is not actually in Aries – it is in Pisces. This is the fault of the Earth's 'wobble' that I have already mentioned. There is a little parable involving religion that might be useful as a mnemonic. At the beginning of historic time, about 10,000 BC, the 'First Point of Aries' was actually in Taurus and this was a time when the Bull featured largely in various religions (I am not necessarily implying any actual causal connection here). It moved into Aries at the time that we get concepts like 'the lamb of God' occurring in religion and then finally Pisces when we get fish appearing as a Christian symbol. Those who can remember the musical 'Hair' will know of a song proclaiming that we are living at the 'dawn of the age of Aquarius' which relates to the movement of the First Point of Aries towards Aquarius.

Equinoxes and Solstices

The Sun is placed at the First Point of Aries at around March 21st and is, as already explained, also simultaneously on the Celestial Equator. This corresponds to the **vernal equinox** – day and night will be of equal length. The other occasion when the Sun crosses the Celestial equator will be on or around 21st September which corresponds to the **autumnal equinox**[17].

---

[15] ignoring our 'totally false but useful' model for the time being and reverting back to the real world where the Earth does actually rotate

[16] in astronomy the term 'meridian' relates to a line that goes through the North Star, the zenith and both due north and due south - the zenith being the point directly overhead

[17] the terms 'vernal' (spring) and 'autumnal' relate to the Northern hemisphere of course

Throughout the period of one year the Sun will move between declinations of +23.5° and -23.5°. The former will correspond to *Midsummer Day* in the Northern Hemisphere, around 21st June, when the number of hours of daylight is at a maximum. This is called the **summer solstice**, although the layperson might be more familiar with the word 'solstice' when it relates to the opposite situation of *Midwinter* around 21st December - the **winter solstice**. At the winter solstice, the Sun is at a declination of -23.5°.

From 21st. December, the day starts getting longer so Christmas is well placed as a midwinter festival, relieving the winter 'gloom' just at the time that it starts to 'go away' (in the sense that the days are getting longer).

As is often said by way of a joke among laypersons, Midsummer Day often bears little relation to real summer. The period around Midsummer Day is rarely the warmest time of the year. The warmest time almost invariably occurs during August - this is an effect caused by the atmosphere and the oceans.

To be slightly pedantic at this stage, Midsummer Day is not the day of latest sunset. The Sun will, for a few days, continue to set about a minute later each day, although it will also rise about 2 minutes later each day. This is due to the fact that the Earth's orbit around the sun is not circular – it is an ellipse. An analogous situation occurs at the Winter Solstice when sunset will get earlier for a few days while sunrise will occur earlier by a greater extent such that the day itself become longer.

Note also that because of its elliptical orbit the Earth will be closer to the Sun during the Northern Winter than during the Northern Summer. It is the Earth's inclination (23.5°) that is responsible for the seasons, most certainly not its relative distance from the Sun[18].

Relative distance to the Sun might be expected to cause some small difference in temperatures, and seasons would be affected slightly as well because the Earth moves slower when it is further out from the Sun. All other things being equal, ignoring atmospheric effects etc. , then a corresponding position in the Southern hemisphere might be expected to have hotter, but shorter, summers than in Britain.

The names of the Tropics have suffered the same fate as the First Point of Aries. Originally the Sun was in Cancer at the summer solstice but (by checking with a planisphere) you can see it has likewise shifted one constellation to the West into Gemini. At this time it will be viewed overhead from the Earth at 23.5° North – the Tropic of Cancer. Likewise at the winter solstice, the Sun will be in Sagittarius, although it was in Capricorn originally, and will be viewed overhead from Earth by a person standing on the Tropic of Capricorn, 23.5° below the Earth's equator.

---

[18] If the Earth's axis was perpendicular to the plane of the Earth's orbit, there would be no seasons and every 24 hours would consist of 12 hours of daylight and 12 hours of night time.

Moon

To consider the motion of the Moon briefly. During a month, the Moon has circled the Celestial Sphere once. Since a month is roughly 28 days[19], then very, very roughly, the moon will move about 15° on the celestial sphere during 24 hours (corresponding to a coordinate shift of about one hour of right ascension).

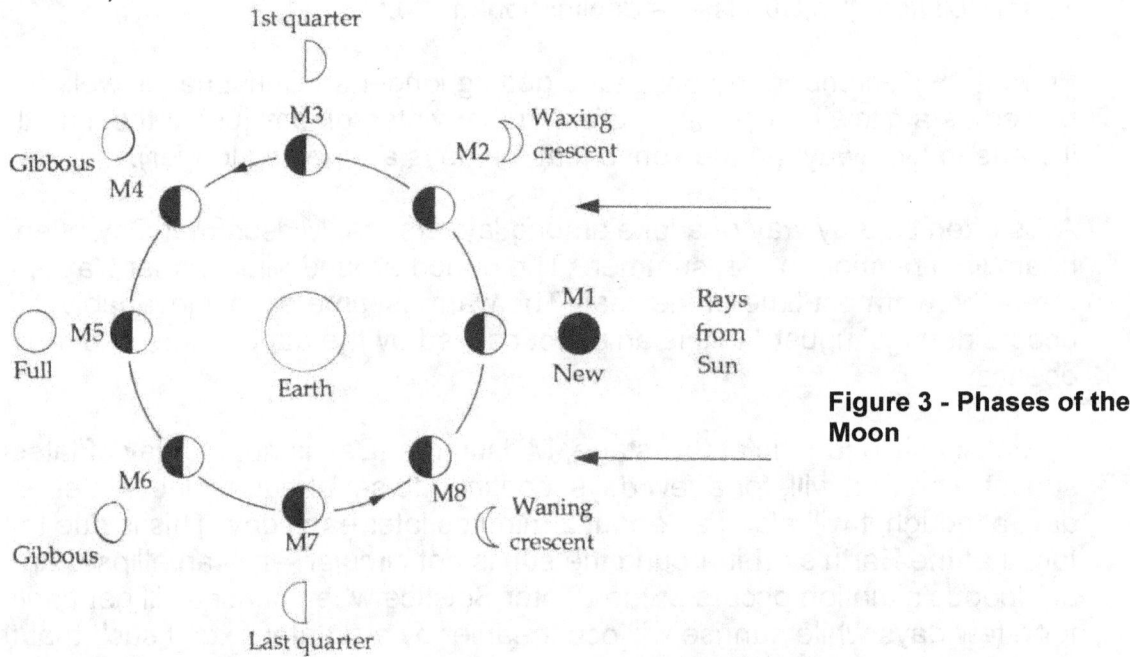

Figure 3 - Phases of the Moon

As it moves on the Celestial Sphere, its phase will alter. Stated in an informal way, whatever is 'on the right' is 'growing'. An illuminated crescent to the 'right' will grow, via first quarter and the gibbous phase (when it is more then half illuminated) to the full moon. When the right of the Moon is dark, this darkness will grow until new moon (when it can't be seen at all) The former situation can be seen in the afternoons and evenings when the Moon is following the Sun down, and the latter situation can be viewed in the mornings when the moon is preceding the Sun across the sky. Refer to the diagram above to clarify the situation (hopefully).

Planets

Venus, Mars and Jupiter can be very bright   Planets out as far as Saturn were known to the ancients, but Uranus was only discovered in 1781. Stars are always point sources of light no matter how powerful the telescope, but a planet presents a disk, which obviously becomes even more noticeable under magnification. Venus can only get about 45° from the Sun, corresponding to about three hours of time – it is either the 'Morning Star' or the 'Evening Star'. Mercury stays even closer to the Sun, and both Mercury and Venus are exempt from the remarks about planetary motion in the next section.

---

[19] This is a very rough figure. I will discuss the situation in more detail later on

Planetary Motion

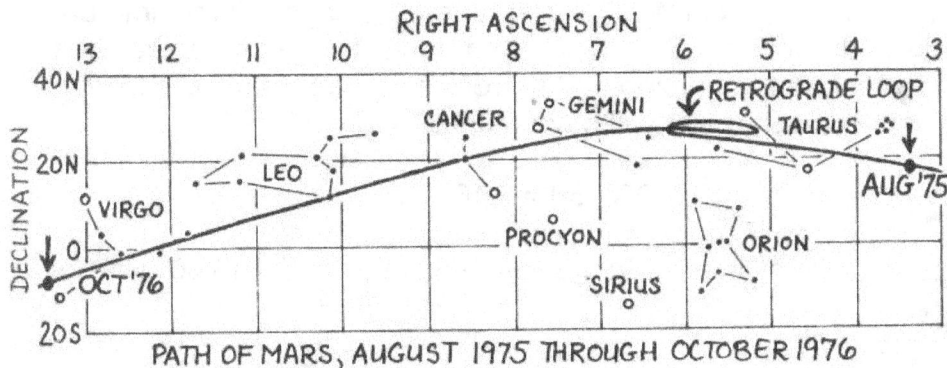

Figure 4 - Retrograde Motion of a Planet

Whereas the Sun and Moon follow a monotonic path, i.e. they always move in the same direction relative to the Celestial Sphere (in an Eastern direction as it happens), the planets can briefly 'change direction'. This is what is attempted to be shown in the diagram above. To understand this phenomenon requires a bit of imagination in a 'thought experiment'. If you were in a car or some vehicle going round in circles accompanied by similar vehicles all keeping to their own rigid circular path. If you were to relate all movement solely to yourself[20] then can you imagine that as you overtake vehicles, they would for a time appear to be moving backwards relative to you. This is exactly what is happening during the retrograde motion of a planet. It is this 'weird' motion that lead the ancients to give them the name of the planets. or 'wanderers'.

Circumpolar Stars, and Field of View

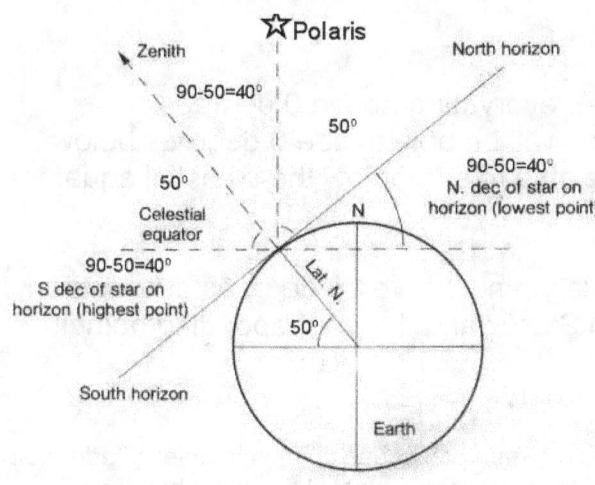

Figure 5

It has been implicit in what I have said so far that the sky will present a different aspect during the course of a year – some stars are only visible in the NIGHT sky for part of the year, and although other stars are visible in the NIGHT sky all year round (the **circumpolar** stars) these will change their attitude with respect to us during the course of a year (i.e. their attitude to us, not to each other, of course, because on our model the stars do not move relative to each other).

---

[20] In a similar way that you can view an adjacent train in a station to be moving <u>relative to yourself</u>, irrespective of which train is actually moving in reality.

Figure 5 attempts to quantify this[21], although the explanation might require some 'real thought'. Viewed from a latitude on Earth of 50° N, Polaris will always have an altitude of 50°. A star 50° away from Polaris (i.e. circling around Polaris during 24 hours) will just scrape the horizon at the time of its lowest altitude in the North but will never approach the horizon in the South – it will always be above the horizon and therefore always visible in the night sky, all year round. Such a star will have a declination of

$$90° - 50° = +40°$$

And thus this star and all stars with a declination greater than +40° will be circumpolar.

By the symmetry of the geometry (you don't need to understand the details if you are mathematically-challenged), from the same latitude, you will be able to see stars down to a declination of -40° (i.e. 40° below the Celestial Equator).

So to summarize the situation:- from a latitude of 50°

- stars of +40 to + 90 will be circumpolar
- stars between -40° and + 40° will be visible, but only for part of the year

In general, you have the 'symmetric' situation that from an Earth latitude of *x* degrees

- Stars above *(90-x)* degrees declination will be circumpolar

- We will be able to see *(90-x)* degrees below the celestial equator when looking south

Consider the two 'extreme' situations

At the North Pole (90 degrees latitude), everything above 0 degrees declination (will be circumpolar and you will be able to see 0 degrees below the celestial equator (i.e. you won't be able to see below the celestial equator at all)

At the equator (0 degrees latitude), everything above 90 degrees declination will be circumpolar (i.e. only the North Star[22]; and the corresponding point in

---

[21] The location of the Celestial Equator on this diagram can cause confusion unless I add a bit more information. Strictly speaking the line representing the Celestial Equator should go thru the center of the Earth but has been moved in a parallel fashion to make the angles clearer. This is in a fashion very
similar to the way that rays of light from the Sun are considered to arrive parallel at the Earth, i.e. the Sun is so far away that the light rays can be considered to be parallel at any point on Earth.
[22] this is obviously using our simple model where the North Star is exactly at the celestial pole

the southern sky) and you will be able to see 90 degrees below the equator (i.e. all the sky will be visible sometime, over the course of a year).

Constellations

If we can now finally move on to the fixed objects on the Celestial Sphere itself –the stars.

It is a burden of history that these stars have been grouped together into constellations, most of which bear no relationship in shape or form to the creature or person they are supposed to be representing. The situation became worse when the Southern sky was treated in like manner and the astronomers responsible made no real pretense of constructing meaningful shapes.

Before considering any specific constellation, I should say a few words about the classification of stars within each constellation. As a rule, the best known stars are designated by a Greek letter, e.g. Alpha Centauri. This was a rule introduced by Johann Bayer in 1603. His method typically was to group the stars into first magnitude, second magnitude etc. and then assign letters within each grouping without any further consideration of relative brightness. In 1603, this further consideration would not have been possible. Consequently, there is no absolute rule that a star designated 'Alpha (α)' will be the brightest star in the constellation. Stars using the initial letters of the alphabet do tend to be the brightest stars in the constellation but this is not a cast-iron rule and there are exceptions.

After using up the Greek letters, Bayer (for some perverse reason) used A (capital letter) followed by b – z (small letters). This provided enough names for his purposes.

This system was extended by later observers using capital letters (B onwards) although the furthest this classification reached was Q (although see the section on variable stars for a description of how this system was picked up again).

An alternative form of classification more convenient when mapping a larger number of stars was the Flamsteed Number. This just assigned a number to stars starting from one edge of the constellation and moving systematically to the other edge.

Anyway, rather than learning the position of objects in the sky via their declination and right ascension, the best way to start off is to learn the more 'recognizable' stars or constellations etc., and then place other objects relative to these 'elementary' constellations.

Two constellations recommend themselves to the beginner-

- **Ursa Major** (the Great Bear), a circumpolar constellation
- **Orion** which is only in the sky in winter.

Ursa Major

Ursa Major is so famous that it will probably need no introduction even to the casual observer or layperson. Known traditionally in Britain as **The Plough** (although strictly speaking the Plough is just the seven main stars whereas Ursa Major is the name of the entire constellation), and also as **The (Big) Dipper,** although this latter name is apparently of American origin[23]. It is also known to the Hobbits as the 'Sickle' and it can be found in Shakespeare as 'Charles Wain', a wain being a wagon and Charles usually assumed to be Charlemagne.

We need to be more circumspect in astronomy and fully realize the difference between the actual constellation of Ursa Major and the designation of 'The Plough' etc. The latter designation is called an **asterism** – it is just being used to reference a small number of stars within the actual constellation itself.

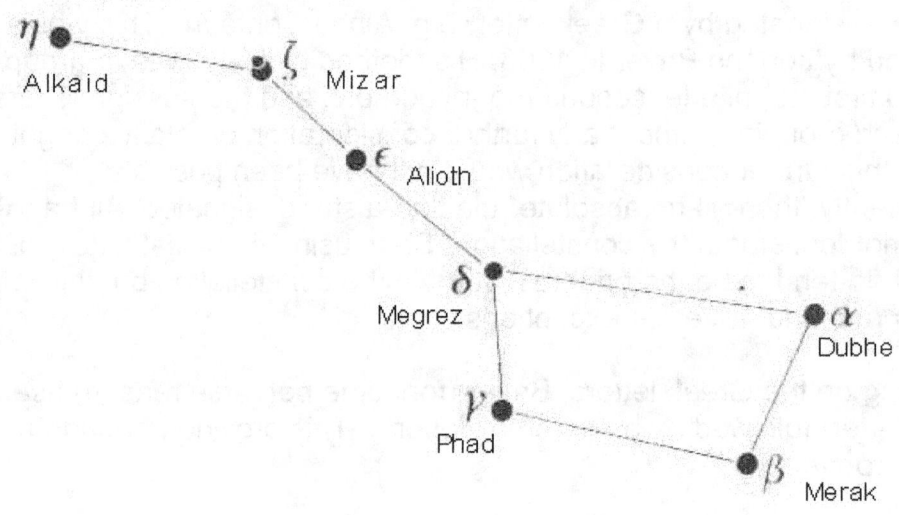

Figure 6 - The Plough

The classification introduced by Bayer can be perceived clearly in the Plough where the stars are of roughly the same magnitude and are listed 'alphabetically' from west to east, although they definitely do not steadily decrease in brightness as you move along the shape. I will talk more about magnitude later, but just accept at the moment that it is a measure of brightness.

Mizar is an especially interesting star. First off it is an obvious **optical** double with Alcor – by optical double, I mean that the stars appear together purely because of the line of sight from us. This distinguishes it from the type of double star where the two stars rotate around each other – a **binary** star. However, further inspection showed Mizar itself to be a double in its own

---

[23] It is indeed a ladle, as you probably have already guessed

right[24] - in this case it is actually a binary, believed to be the first binary to be detected telescopically[25]. And, yet further, each of its component stars is a **spectroscopic binary**, i.e. two stars orbiting around each other, whose binary nature was only detected via its Doppler shift. In fact, Mizar A[26] was the first spectroscopic binary to be detected – in 1889 by E.C. Pickering

If I can now put the Plough within its context in the sky. The line joining Merak and Dubhe will point towards the North Star (in Ursa Minor), although it is not necessarily as easy as might be generally suggested. Ursa Minor and Polaris are definitely fainter than Ursa Major.

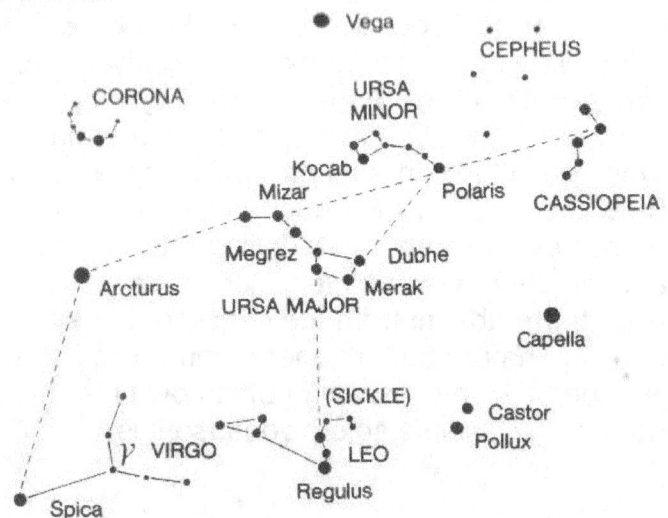

Figure 7 - Ursa Major and its environs

Less well-known is the fact that if you follow the tail of the Plough downwards in an arc, you come to Arcturus (in Bootes), the fourth brightest star in the sky and the second brightest star to be seen from Britain.   It gets its name because it is the 'bear watcher'. Following the arc further down leads you to Spica, a first magnitude star in Virgo.

Adjacent to the tail of Ursa Major is the constellation of Canes Venatici – a small constellation but that's one more to add to your knowledge anyway. 'Below' the Plough, you will find Leo, and you can perceive from the diagram the adjacent zodiacal constellations of Virgo and Gemini .

Over on the opposite side of Polaris, you will see the W-shape of Cassiopeia[27].

---

[24] i.e. separate to its claims to be a double with Alcor
[25] by Riccioli in 1650
[26] The two original components being classified as Mizar A and Mizar B
[27] or 'M' shape, if you like

Orion

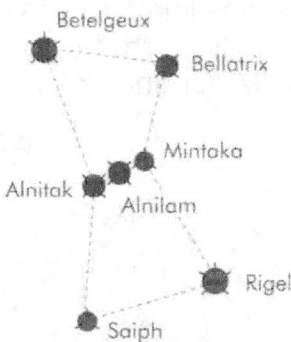

Figure 8- Orion

Turning our attention now to Orion – when it is in the sky, namely in Winter, you should have no difficulty in locating it. It contains 10% of the 70 brightest stars in the sky. α-Orionis is Betelgeuse although Rigel is brighter than it. Betelgeuse is a red **supergiant** (see later) which can be noticed clearly in a telescope but to the naked eye it might tend to appear as reddy-white, since all faint light sources are interpreted by the human eye as being white. Rigel is also a supergiant, but of a white color. The three stars at the center form Orion's Belt. Below Orion's Belt is the Orion Nebula which is best viewed using averted vision (i.e. by not looking at it directly), a technique which relies on the fact that rods (which detect faint light) are absent from the center of the eye. This nebula is actually quite complex in nature but in general you could classify it as an *emission nebula*, a nebula which is giving off its own radiation as opposed to a *reflection nebula* which is visible solely because it is reflecting light.

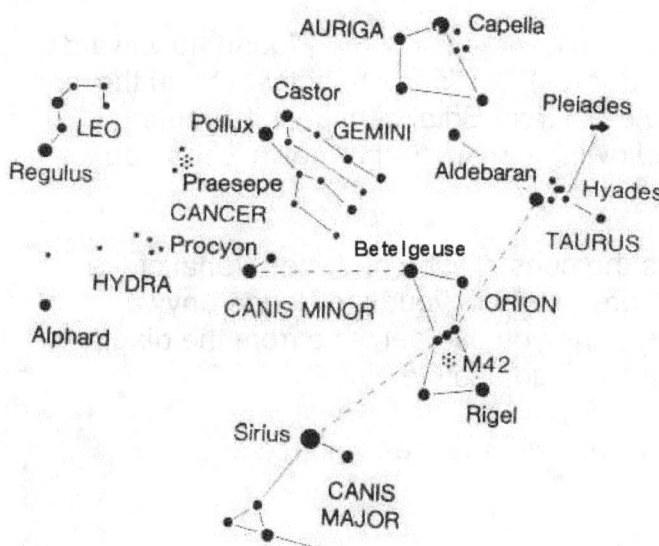

Figure 9 - Orion and its environs

Looking at the constellation in its wider context, we can see straightaway that following Orion's belt downwards will lead to Sirius[28] in Canis Major, the brightest star in the sky. Following the belt upwards will lead to the orange star Aldebaran in Taurus – this is closely connected with the Hyades[29], an **open cluster** whose brightest stars form a V-shape. A bit further on is the Pleiades (M45), another open cluster (also called a **galactic cluster**). Like all nebulas, this

---

[28] Sirius, the Dog Star, rises just before the Sun during the Summer. Since it was once thought that the Dog Star and the Sun combined to produced the heat, the hotter days of Summer are sometimes called 'Dog Days'. To the ancient Egyptians, Sirius heralded the start of the Nile flood. Its name means 'trembling' or 'sparkling' – its twinkle is strong because it is so bright and because it is low down meaning its light travels thru a thicker layer of atmosphere, which can also introduce coloring.
[29] not within the field of 7 x 50 binoculars

is an object that can be seen best using 'averted vision'. Using an optical instrument, the naked-eye 'smudge' will become a beautiful object to observe It has about 250 stars of which the brightest nine are named after the seven Pleiades of Greek mythology and their parents, and has much dust producing a **reflection nebula**.

Although it is not shown on the diagram, following the 'line' further up you come to Perseus, notable for its variable star Algol - it is actual a binary star which dips in brightness when one component goes behind the other.

To the upper left of Orion you have Castor and Pollux and below them, Procyon. Although sky pollution is the bane of an astronomer's existence, these objects will be obvious in a polluted sky where they are likely to be the only objects visible in the said positions. Directly above Orion you have the pentagon of Auriga, or more particularly its red star Capella.

Figure 9 also includes certain **Messier objects** although M42 (the Orion Nebula) is the only one explicitly mentioned as such. This list of objects was originally compiled by Messier as objects that 'were not comets' for the benefit of searchers of these latter objects. Nowadays, there are 110 Messier objects altogether and, loosely speaking, they constitute the most easily-seen of non-stellar objects. They include a wide-range of objects – for example, those included in Figure 9 - Orion and its environsare

- M42, the Orion Nebula – I have already described this as an emission nebula. More specifically, an important constituent is a type of emission nebula called an **HII region**, a region of star birth. The stars are usually hidden within the gas clouds, but in M42 several young stars can be viewed, forming what is known as the Trapezium. (It is HII regions and bright stars that delineate the spiral arms in a spiral galaxy.)

- M45, the Pleiades – an open (or galactic) cluster

- M44, Praesepe – an open cluster

Figure 10 - Leo

Leo

Leo is a distinctive constellation once you have located it for the first time. Its brightest star is Regulus and its second brightest is Denebola, whose name stems from the Arabic for 'tail'. I emphasize this because the 'tail star' in Cygnus is called Deneb for the same reasons. γ is Algieba and is a double star of the type sometimes known as 'headlights' – the two yellowish components are of almost equal magnitude and of fairly high magnitude..

Pegasus, Andromeda, Cassiopeia, Perseus

Perhaps Pegasus should have been mentioned earlier because it typifies one of the problems for beginners working with star maps. Despite its seemingly straightforward appearance on the planisphere or on a star map, Pegasus is a notoriously hard object to locate with certainty. This is because a beginner might be uncertain of its size on the sky for one thing. The second problem might be in recognizing how the square is slanted with respect to the Earth's horizon. The sides of Pegasus might be parallel to the coordinate lines but these coordinate lines are slanted on the sky.

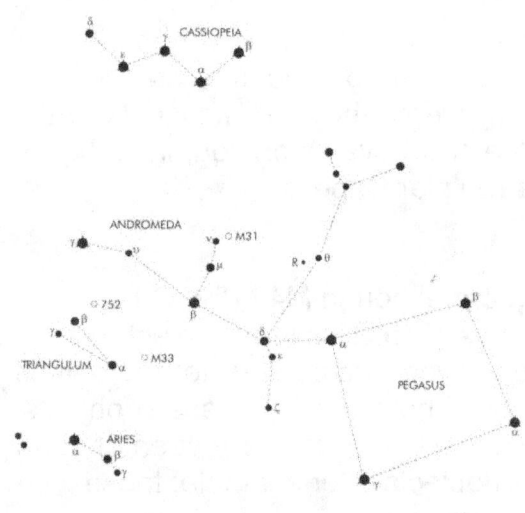

Figure 11 - Square of Pegasus, Andromeda, Cassiopeia

The main problem is the size – it is useful to know that the Sun (and Moon) subtend an angle of ½ °, so the stars of the Square of Pegasus, at about 7 degrees apart, will be about 15 Sun diameters apart.

Leading off from the top-right corner of Pegasus is Andromeda. In fact is an oft-mentioned quirk that the top-left star in the Square is actually α-Andromedae. This and two other bright stars arranged almost in a line form the most visible representation of the constellation. Just up from the 'middle' star lies M31, the Andromeda Galaxy[30]. This is classed as a naked-eye object although it is by no means an obvious object. It will be a misty ellipse a couple of degrees across with no spiral structure. Even with optical aids it is not immediately obvious to the beginner that you are looking at the right object. α-Andromedae is an impressive double star composed of a second magnitude orange and fifth magnitude blue-green components.

In Cassiopeia α and β point to M52, an open cluster.

It is not shown on the diagram, but Perseus lies just off the diagram to the top left. The Double Cluster in Perseus is an 'omission' from the Messier List and is an open cluster lying between Perseus and Cassiopeia. They have star names h and χ. They are young clusters and there is a debate as to whether they are connected in any way.

---

[30] commonly described as the furthest object seen with the naked eye – complete with its satellite galaxy M33 - at about 2.5 million light years.

If you know Greek mythology, you will known that Cassiopeia was the mother of Andromeda. The latter was saved from death by Perseus who rode into the rescue riding the flying horse Pegasus. Cepheus was Andromeda's father and Cetus (further South) was the sea monster to which Andromeda was to have been 'sacrificed' before Perseus arrived on the scene.

Summer Triangle

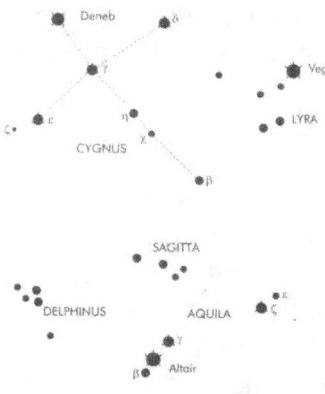

Figure 12 - the Summer Triangle

The Summer Triangle is a very prominent feature during the season in question. It is formed by

- <u>Vega</u> (Lyra) almost overhead,
- <u>Deneb</u> (Cygnus), a very bright star which is also very far away. It is a supergiant.
- <u>Aquila</u> (Altair)

Cygnus itself forms what is often called the 'Northern Cross'. It differs from the Southern Cross in having a star at its 'intersection'. α-Cygni is Deneb which is the nineteenth-brightest star but is notable because Deneb is a white supergiant and far away – most other bright stars are bright because they are 'close'. β-Cygni is Albireo – a double star, with an amber giant (mag 3.1) and blue-green (mag 5.1) components which can be seen without a large telescope.

Cygnus lies in the Milky Way (which will not be obvious from polluted skies) – Altair lies just on the 'lower edge' of the Milky Way with Vega 'above it'. The region from Cygnus to Sagittarius roughly corresponds to the 'Summer Milky Way'. There are great dark rifts in the Milky Way through the constellations of Cygnus and Aquila.

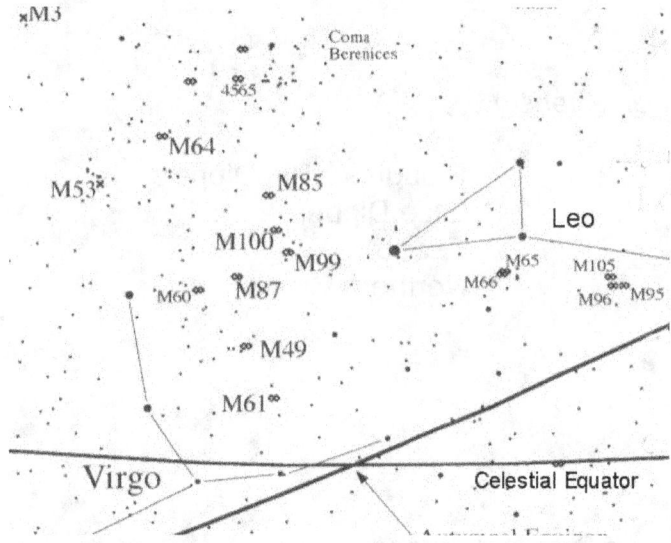

Figure 13 - Virgo

Virgo

Virgo is actually shaped like a 'Y', forming a 'bowl' which contains many objects from the Virgo Cluster, a cluster of galaxies adjacent to the Local Group of which the Milky Way is a member. You can see from the diagram the preponderance of Messier objects. This Y-shape is even more pronounced if you include Denebola from Leo, although forming shapes that transcend the traditional constellations is officially 'frowned upon'. The star at the bottom of the 'Y' is Spica and the star at the intersection of the 'stem' and 'bowl' is Porrima (γ), a well-known binary star.

Sagittarius and Scorpius

Sagittarius and Scorpius are low down from Britain but can be good objects to observe during Summer from the Mediterranean regions.

Scorpius[31] is traditionally considered to be one constellation that can indeed be pictured as what it is actually intended to represent. The red star Antares (carrying an allusion to the planet Mars[32]) is a very distinctive object. Only the head of Scorpius is potentially visible from Britain.

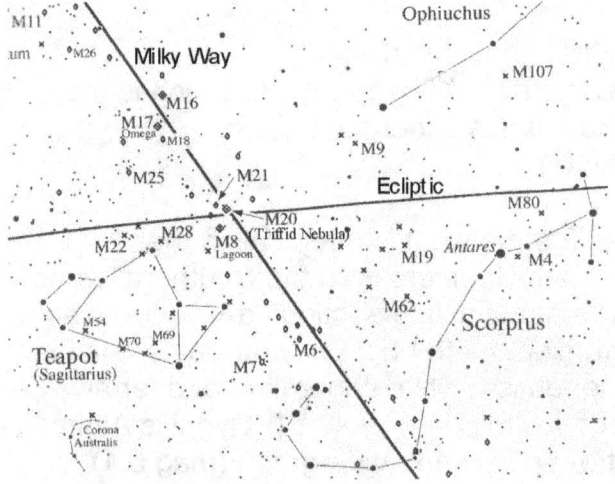

According to one legend, Scorpius is the scorpion that stung Orion to death. They stand at opposite ends of the sky such that for most of the time they are not in the sky together.

Sagittarius, nominally an Archer or Centaur, could be 'visualized' as firing his arrow at the Scorpion.

Figure 14 - Sagittarius and Scorpius

Analogous to Ursa Major, it is often referred to by an asterism formed a few of its stars : the Teapot'[33] in which form it is shown in the diagram. A group of stars near the spout are called the Head of Steam.

The center of our own galaxy is in Sagittarius although it is hidden from view by dust

Constellations sometimes known by an asterism

Great Bear	Plough, or Big Dipper
Little Bear	Little Dipper
Sagittarius	Teapot
Cygnus	Northern Cross

---

[31] Note the name. The constellation is called Scorpio in astrology
[32] Mars in Roman mythology is equivalent to Ares in Greek mythology
[33] Although note what I said about Ursa Major : the Teapot is only a part of the wider Sagittarius constellation.

<u>*Constellations similar to a letter*</u>

Cassiopeia	W
Virgo	Y
Taurus	V

<u>*General description of some constellations*</u>

Aries	Three Stars
Canes Venatici	Two Stars
Auriga	Pentagon
Cepheus	Pentagon
Pisces	Circlet (of 5 stars)
Delphinus	Kite
Corona Borealis	Semi-Circle
Pegasus	Square
Hercules	Keystone
Vega	Parallelogram
Corvus	Trapezoid/Tilted Kite/Sail
Bootes	Kite

If I can just briefly mention a few odd topics to finish off this chapter

Magnitude

As far back as the 2nd century BC, Hipparchus classified stars into six classes of magnitude, with first magnitude being the brightest. Today we can differentiate brightnesses within each class, so that a second magnitude star could have a magnitude value anywhere between 1.5 and 2.5. In 1856, the system was systemized by N.R. Pogdon so that an exactly first magnitude star is 100 times as luminous as an exactly sixth magnitude star[34] - that is 100 times as bright when a scientific instrument is used to measure its magnitude. The human eye (and brain) work differently - we do not detect magnitudes 'linearly' but logarithmically, so that each increase in magnitude corresponds to an increase in magnitude of

$$\sqrt[5]{100} = 2.51$$

Hearing operates in a similar way in that, for example, the difference in intensity between 4 and 5 decibels is 10 times as large as the difference between 3 and 4 decibels.

With the help of telescopes we can obviously see objects much fainter than sixth magnitude.

---

[34] i.e. stars of magnitudes 1.0 and 6.0.

Sidereal and Synodic time periods.

The **sidereal periods** of a planet's orbit is officially defined as being the period calculated in relation to the fixed stars but on a more basic level is just the planet's 'inherent' orbital period.

This would be different from the **synodical period** which would be the time that we on Earth would perceive the planet takes to attain the same position in the sky.

Taking Mars as an example

- its sidereal period is 687 days, i.e. the time it takes to circle the Sun.
- its synodic period is 780 days, for example the time between two oppositions[35] of the planet

The differences are obviously caused by the fact that the Earth is moving as well.

Take the Moon as another example. The Moon completes one orbit around the Earth in 27.3 days – this is the sidereal period. This is not what we would commonly conceive as the period of the Moon – we would consider something like the period from one full moon to the next full moon to constitute its period, which is the synodical period[36] of length 29.5 days.

You can see from the diagram what is happening. During the 27.3 days that the moon takes to orbit the Earth, the Earth has moved along its own orbit about the Sun. The moon requires another 2 days or so until it can attain a Full Moon.

The further and further out you go, the closer and closer the synodical period comes to 365 days, and this is despite the sidereal period becoming ever larger (Neptune orbits in 165 years for example). It is precisely this long sidereal period which accounts for the synodical period being close to 365 days. Within an Earth's year, the planet has moved so little with respect to the Celestial Sphere, that the earth only needs to 'move' a bit further (a few days) to line up the planet in the same position as it was previously.

---

[35] A planet is at opposition when it is directly opposite the direction of the Sun in the night sky
[36] The synodical period refers to how we would see it, not necessarily from full moon to full moon-the synodical period could also be from new moon to new moon or any other valid period

Snippet

- The Gregorian calendar which we use in everyday life is the only calendar to keep in synch with the Earth's orbit round the Sun and therefore also the seasons. All other calendars tend to be based on the Moon and are therefore 'out of synch', so events like Chinese New Year, Ramadan etc. occur at slightly different (Gregorian) dates year after year.

Chapter 2   Stars, Part 1

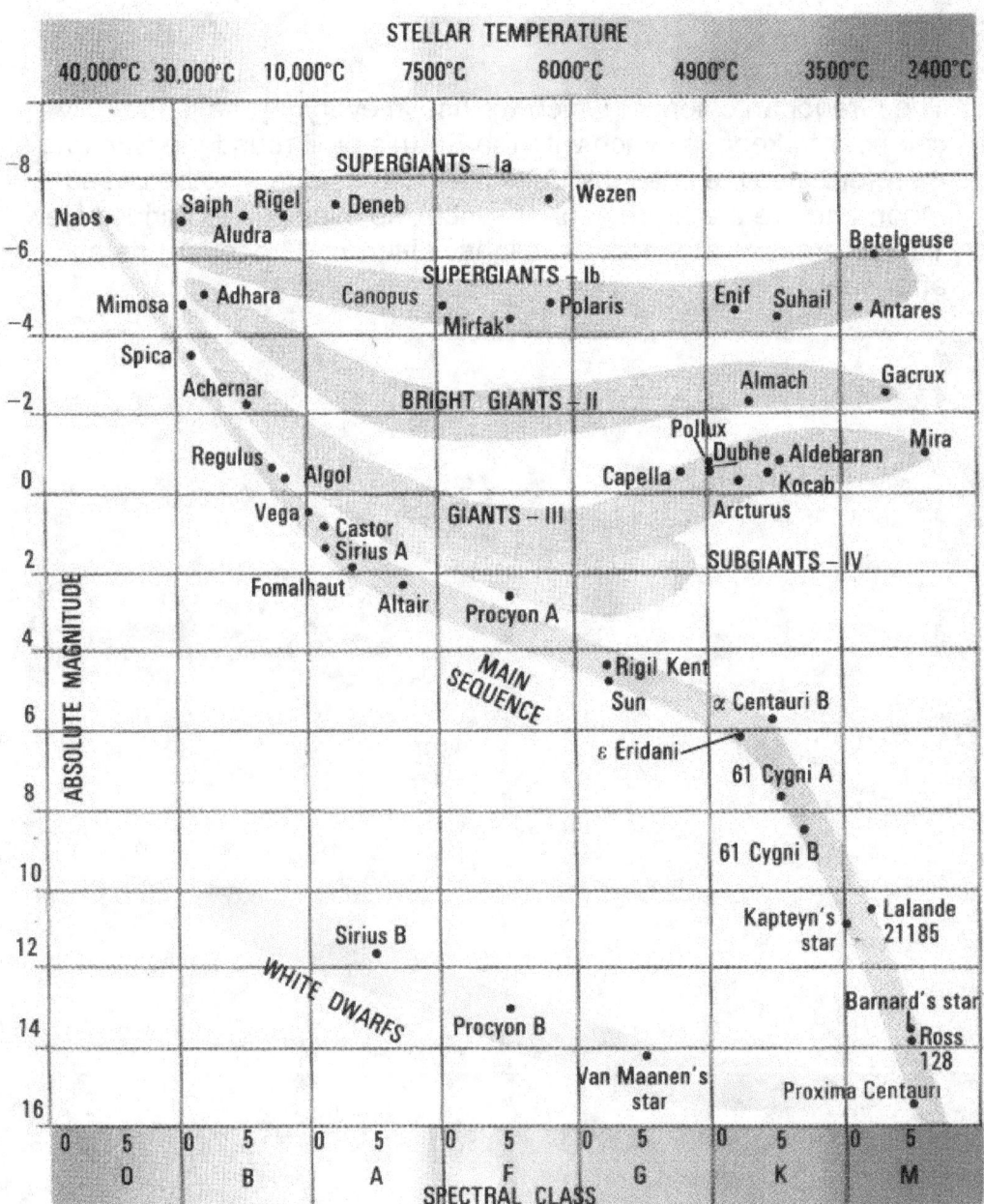

Figure 15 - Hertzsprung Russell (HR) diagram

The Hertzsprung-Russell Diagram (above) essentially relates the inherent luminosity of a star to its surface temperature.

If I can just expand on what is represented on the axes.

The vertical axis can be seen to represent the **absolute magnitude**, which is different from the **apparent magnitude** mentioned up to now. The absolute magnitude is the magnitude a star would have if it was viewed at a distance of 10 parsecs[37] (a parsec is about 3.16 light years).

The vertical axis could just as easily have represented the **absolute luminosity**, i.e. the inherent luminosity, as opposed to the luminosity that we ourselves detect on Earth (which would be affected by the distance of the star). The luminosity is measured in Watts, so it is just the power of the star's emissions

The horizontal axis can likewise represent two equivalent quantities

- the **surface temperature** of the star. Notice that the scale is marked differently to what you would expect – the temperature increases towards the left. Here the temperature is given in degrees Celsius /Centigrade but often it is measured in Kelvin (K) – the Kelvin scale is graduated identically to the Celsius/Centigrade system except that the Kelvin scale starts at absolute zero, so

$$0 \text{ K} = -273° \text{ C}$$
$$273 \text{ K} = 0° \text{ C} .$$
$$373 \text{ K} = 100° \text{ C}$$

and so on

- the **spectral class**, which are now ordered as OBAFGKM (Oh Be A Fine Girl Kiss Me). In the original classification drawn up at the end of the 19th century, they were given a straightforward alphabetical sequence, from the simplest-looking to the more complex, except that the classes had to be re-ordered the more they learnt about spectra.

There is another feature which I need to discuss before looking at the HR Diagram in more detail, namely the manner in which color is dependent on temperature.

We introduce a theoretical entity called a Black Body[38] - this is a body that absorbs all radiation incident on it. Normally objects in everyday life do reflect light of particular frequencies and this determines their color as we perceive them, but a body that reflects no radiation would be black. This is in line with our experience in everyday life whereby black objects will heat up in the Sun faster than objects of other colors by virtue of the fact they are reflecting less radiation than other colors.. The concept of a Black Body takes this to the extreme and considers a body that reflects no radiation at all.

So our theoretical Black Body will not reflect light but will heat up due to the absorbed radiation and start to transmit its own 'inherent' radiation. This radiation will be characteristic of the temperature of the Black Body.

---

[37] An object which had a parallax of one second of arc would be at a distance of one parsec (explanation of terms will be given later)
[38] Not to be confused with a 'Black Hole'

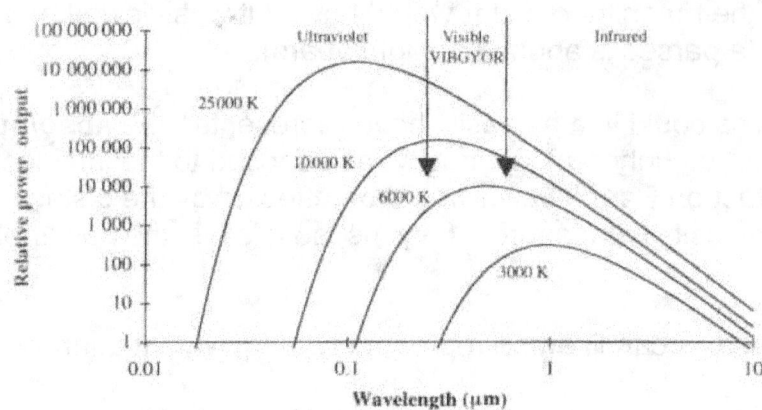

Figure 16 - Black Body Radiation

Figure 16 - Black Body Radiationshows the characteristic 'Rock of Gibraltar' shape of a Black Body curve. Consider the curve corresponding to 3000K – if the body was heated further then two things happen

1. the peak of the curve shifts – to lower wavelengths (or equivalently, to higher frequencies[39])
2. The total energy emitted increases, as represented by the larger area under the curve

In the figure, the 3000K body is peaking in the infra-red, but the 6000K body is peaking at the blue end of the visible light range.

The general features can be illustrated at least qualitatively by thinking about a domestic device like an electric oven hob despite the emission probably not being black-body. As we switch it on, the peak of the curve will shift to lower wavelengths and we will soon detect the Infra-Red radiation – as it heats further the peak enters the visible-light range, at the lower-energy red end.[40]

If we had a device that we could heat even further, the peak of the curve would shift to 'higher-energy' colors, expect that as it does this it will also be emitting appreciable 'amounts' of other colors and will thus tend to appear white[41] - you might have heard of phrases like 'the white-heat of technology' when referring to industry (also by analogy, the temperature of industrial furnaces can be measured from their colors).

The reason why I talking about black-body radiation is that this radiation is characteristic of the radiation given out by a star.

---

[39] Although the diagram shows wavelength, it is really the frequency that determines the nature of electromagnetic radiation – what type of radiation it is, what color it is, etc.
[40] From Einstein's equation $E=hf$, where E is energy, h is Planck's Constant and f is frequency, the higher the frequency, the higher the energy (and vice versa)
[41] While light being a mixture of all colors. Think of a prism which can take white light and split it into its constituent 'rainbow' colors

Stars of even higher temperature than that already mentioned and peaking at the high-energy' blue end of the visual spectrum will also tend to look white, but there are quite definitely blue stars to be viewed in the sky[42].

Note that the Black Body radiation for 6000K (approximately the surface temperature of the Sun) peaks in the middle of the Visual Spectrum. For the reasons already given, all other things being equal the Sun would appear to be whitish. The atmosphere strips off its blue emissions preferentially, producing the blue color of the sky, leaving the Sun a bit 'more yellow'. This affect can be exaggerated further at Sunset when all further colors are scattered by the atmosphere leaving the Sun a red color.

Anyway, back to the HR diagram you can see the relevant colors indicated on the diagram – red stars to 'the right' and bluish stars to 'the left'.

Main Features of the Diagram

There is an obvious quasi-straight line on the diagram called the **Main Sequence.** Stars on the Main Sequence are hydrogen-burning stars and once a star settles on the Main Sequence it effectively stays in the same place for the duration of its hydrogen-burning phase (to a first approximation). The mass of these stars vary from 0.05 solar masses up to probably about 50 solar masses.

Other main features of the HR diagram to note at this stage are

- **Giants**, particularly those at the low-temperature end to the 'right' – the Red Giants. The large size of these Red Giants ensure that their total luminosity is high despite the fact that their surface temperature is low.

- **Supergiants**, which extend more obviously across all colors (or equivalently temperatures). As we will be discussing, Giants and Supergiants are totally different things.

- **White Dwarfs**, objects that paradoxically are very hot but inherently quite dim - the main paradox being how something so small can be so hot.

Protostars

I could start right at the very beginning and talk very briefly about the dense[43] molecular gas clouds from which stars are born. These are colder than other regions of the *Interstellar Medium*. If such a cloud should collapse due to its low temperature and 'high density', potential energy will be converted into heat and this heat is eventually high enough, without any nuclear fusion,

---

[42] Andromeda

[43] dense in the astronomical sense – they would be described as perfect vacuums if they were reproduced in the laboratory

to produce a 'proto-star'[44] It will 'wander around' to the right of the main sequence before actually settling down on the main sequence when the core becomes hot enough, and pressure is high enough, to initiate hydrogen fusion. Before entering the main sequence it will probably pass thru the T Tauri stage (if it has a mass between 0.2 and 2 solar masses), a type of star notable for a strong stellar wind. If the Sun went thru such a T Tauri stage (as is suspected), the wind would have been capable of expelling much of the primordial gas from the Solar System, including any hydrogen atmospheres possessed by the terrestrial planets. Any subsequent atmosphere on these planets

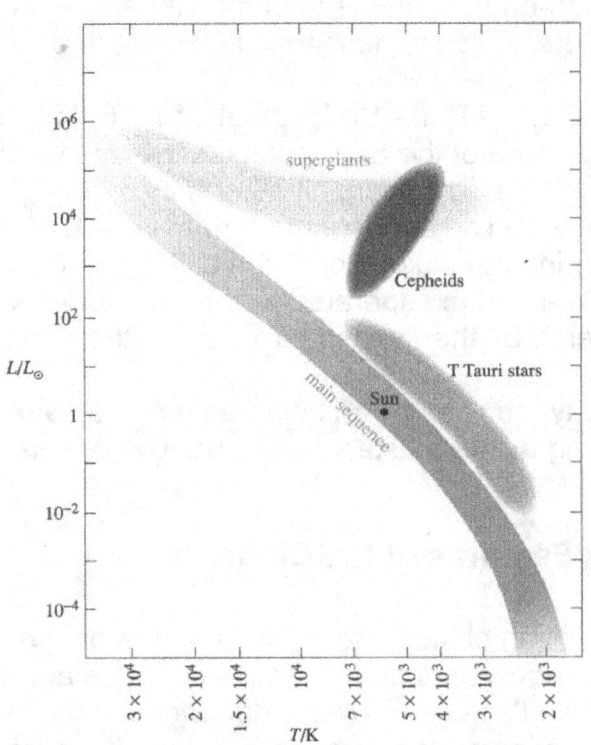

Figure 17 – HR diagram showing early T-Tauri stars (irregular variables) and Cepheids (regular variables)

has come from outgassing from the interior of the planet itself. The wind could also have 'stunted' the growth of the gas giants by removing the gas that they might otherwise have accreted[45].

Main Sequence

Main sequence stars are hydrogen burning stars. Stars typically spend 90% of their lives on the main sequence. They will occupy a position related to their mass – the higher the mass of a star the higher the point on the main sequence that it will occupy. Now the higher the mass of a star, the more efficiently it will be using its fuel, and consequently the shorter its life-time (For the Sun, the core is at about 15 million degrees - although the surface is 'only' 5800 degrees- and it will have an estimated life of 10 billion years[46])

A markedly greater number of lower mass stars are formed than higher mass stars, although the faintness of the latter will mean they are less 'noticed' (obviously). Stars at the bottom right of the Main Sequence are called **red**

---

[44] I will call it a proto star if it is definitely on its way to becoming a star, as opposed to a hot gas cloud that is not going to end up as a star.
[45] The further out a planet, the more slowly it will accrete gas - such that if there were no T Tauri wind, the relative sizes of the gas giants would be noticeably different (under one current popular theory anyway).
[46] and its current age is 5 billion years, half of its estimated total lifetime

dwarfs[47]. One of these, at least, is well known – Barnard's Star, which has the largest proper motion of any known star. It moves half a degree (one Moon magnitude) in 175 years.[48]

Stars will 'peel-off' the Main Sequence sequentially starting from the top, and these stars will start moving to the right on the HR diagram.

A *color-magnitude diagram*, a precursor of the full-blown HR diagram, maps data from a single cluster. All the stars in the cluster are assumed to have been born at the same time and a color-magnitude diagram will show this 'peeling-off'. The **turn-off point**, the upper point of stars still on the Main Sequence can be used to calculate the age of the cluster.

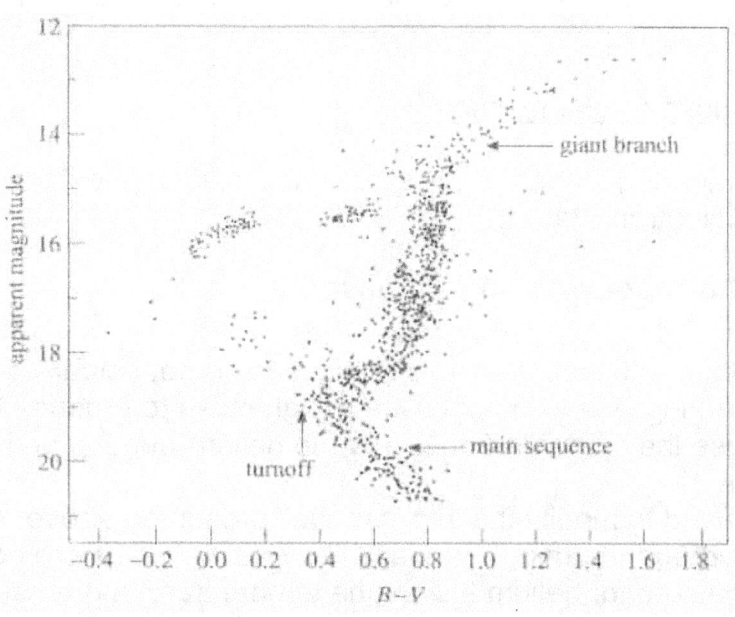

Figure 18 – Color-Magnitude Diagram of a star

After the Main Sequence

Now it is only the core of a star that is producing energy. If we present a 'basic model' as follows :- equilibrium in the star[49] is reached when hydrogen fusion[50] occurs. When hydrogen in the core is depleted and converted into helium, fusion stops and the star starts to collapse. This raises the temperature and pressure in the core such that eventually the helium will start to fuse, producing a new equilibrium. When the helium is depleted, we might imagine the scenario repeating itself – collapse and then a new equilibrium, with heavier and heavier elements being fused in the core.

How stars will behave after the Main Sequence depends on their mass. For simplicity at this stage, I will identify two separate classes - above and below 4 solar masses. (thereby ignoring the behaviour of stars in the 'boundary region').

---

[47] There are 300 stars within 10 parsecs of the Sun, and 70% are red dwarfs

[48] Up to now, we have been working with a 'model' where the stars do not move on the Celestial Sphere. Obviously in the real world they do move but extremely slowly. Barnard's Star moves so 'fast' because it is close.

[49] By equilibrium, I mean the inward pull of gravity is balanced by the outward pressure caused ultimately by the fusion in the core

[50] Hydrogen nuclei join together to form a new nucleus (of helium) and releasing energy

To summarize the progression in each case:

Below 4 solar masses

Main Sequence > Red Giant > Cepheid (maybe) > Planetary Nebula > White Dwarf

Above 4 solar masses[51]

Main Sequence > Supergiant (including Cepheid stage maybe) > Supernova > Neutron Star (or Black Hole)

Stars below 4 solar masses

Stars with less than 4 solar masses do not actually get farther than helium-burning. They do not have enough mass to produce the conditions required to fuse the carbon produced by the helium-burning and therefore start 'to die'.

I need to modify the 'basic model' presented above. At the cessation of hydrogen-burning[52], a star will collapse, but before conditions become ripe in the core for helium fusion, the temperature and pressure in the inner section of the hydrogen envelope become suitable for yet more hydrogen burning – we have '**shell-burning**'.

The effect of shell-burning will be that, although the core still tends to collapse, the envelope will expand. The star will swell to gigantic size, the surface cooling as it does so. The surface changes color to red – and when, in addition, helium core burning starts, we have a full-blown '**Red Giant**'.

Helium burning proceeds by what is called the 'Triple Alpha Process'[53]. Two helium nuclei can fuse to form a Beryllium nucleus but this is unstable and will decay swiftly. If before this decay occurs a third helium nucleus fuses to the beryllium nucleus a stable carbon nucleus will be produced. This feature of helium fusion is a factor to be considered in the production of nuclei in the early Universe, when only hydrogen and helium are in fact produced.

---

[51] the sequence I am describing probably only applies in full to stars above 8 solar masses
[52] By this I mean hydrogen fusion
[53] A helium nucleus is equivalent to an alpha particle
in radioactivity – a nucleus formed of two protons
and two neutrons

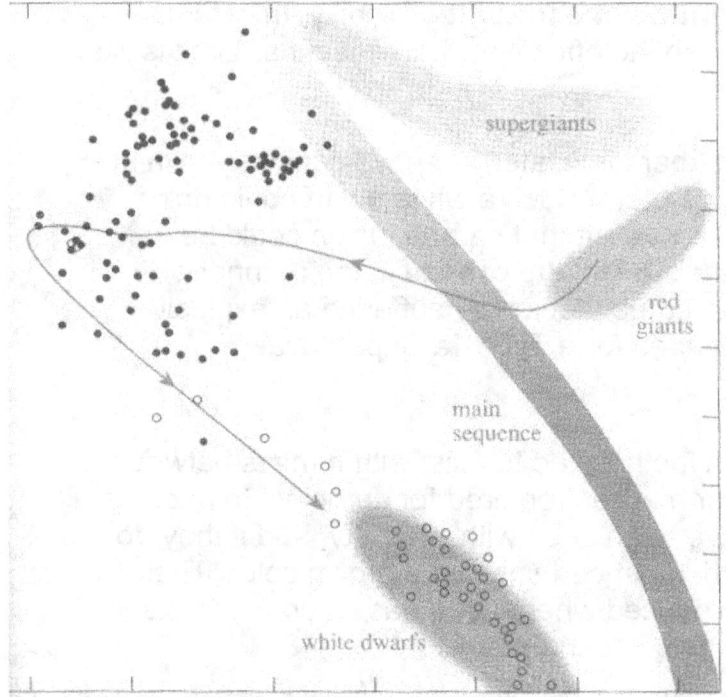

Figure 19 – tracks of post-Main Sequence stars

Examples of how the star will then behave on the HR diagram are shown schematically in Fig 19. Looking at the path annotated 3Mo (3 solar masses) the star's subsequent history is as follows

B-C indicates the onset of hydrogen shell burning
D is the start of helium core burning
F designates the onset of helium shell burning

Eventually the star will drift across to the upper left of the HR diagram. Instabilities in its core regions lead to an expulsion of its outer envelope, producing a **planetary nebula**, so called because its disk-like appearance resembles the appearance of a planet, or did so to 18th century observers anyway. The envelope is moving away at 10-30 km per second. Some planetary nebulae are spherical, some dumbbell shape. The envelope commonly looks like a ring around the star[54] but we can see through to the star only because of the thinness of the gas – this gas is really surrounding the star.

The core will settle down eventually as a **white dwarf** – a very faint object but with a temperature much higher than was to be expected for a small star. It

Figure 20 from Red Giant to White Dwarf – black dots represent planetary nebulas

---

[54] and indeed the Ring Nebula (M57) is a well-known example

was only when quantum mechanics was invented in the 1920s that astronomers came to understand the nature of white dwarfs – they are objects too dense to understand using classical physics.

Friedrich Bessel had indirectly detected the white dwarf Sirius B in 1844 when he was able to detect a jiggle in Sirius's orbit[55]. Sirius B was first seen visually in 1862[56] (it takes 50 years to orbit the main star) but at first they could hardly believe what they were seeing – this was an object which was very small (diameter similar to the Earth's) but which was producing an enormous gravitational pull for its size.

In the early 20th century, Sirius B was calculated to have a mass between 0.75-0.95 solar masses (current value is near the top end), a luminosity 1/360 that of the Sun, but with a spectra indicating a temperature of about 8000K (current value is about 24,000K). The radius was calculated to be about 19,000 km, but this is about three times the currently accepted value (which is similar to the radius of the Earth).

Quantum mechanics was able to explain this by showing that white dwarfs are composed of a highly dense form of matter called 'degenerate matter'. For future reference, I should also state that this degeneracy is related to the electrons – in a very hand-waving description, they are resisting compression in the same way that electrons in an atom resist the temptation for electron levels to be compressed.

In 1930, Chandrasekhar[57] introduced special relativity into the problem and showed that white dwarfs have an upper limit to their mass – usually given as about 1.4 solar masses, the **Chandrasekhar Limit**. In reality, the star is expected to be rotating very fast, so the effective Chandrasekhar Limit is likely to be nearer to two solar masses

We will talk about the Chandrasekhar Limit later when we consider a white dwarf as one element of a binary system. Such a white dwarf could draw matter off its companion to such an extent that carbon fusion could be initiated. Since degenerate matter is unable to expand in the manner of normal plasma, it will react to this carbon fusion by suffering an explosive runaway. This is the favored scenario for a Type 1a Supernova.

Brown Dwarfs

These are faint objects that are hypothesized to exist with a mass between that of a planet and the 0.05 solar masses required for an object to become a hydrogen-burning star. None have been seen with certainty, and if they do exist they will glow with a dull red, most certainly not a brown color. Their radiation comes from the heat produced whenever a gas cloud contracts.

---

[55] He also detected a jiggle in Procyon's orbit, but the Sirius story is better known. Procyon B was detected visually in 1898.
[56] by Alvan Clark. By all accounts he initially thought it was a flaw in the lens he had just made
[57] From India, studied and worked originally at Cambridge University. Moved to Chicago University in 1937. Nobel Prize winner 1983. The Chandra X-ray space probe of 1998 is named after him.

Chapter 3   Stars, Part 2

Now the more massive stars can indeed initiate fusion of carbon and beyond. We need to note however that iron forms a 'limit'. It is not possible to fuse iron to produce energy - fusion of iron requires an input of energy. Once it reaches this stage, the star is dying and, according to the theory, will go supernova

Figure 19- Rest energy per nucleon w.r.t mass number

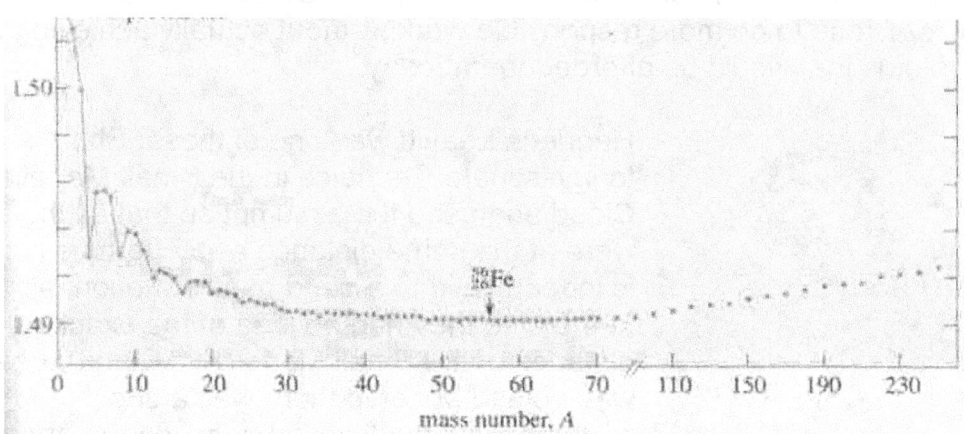

The above figure shows that you can carry fusion as far as iron with production of heat, but no further. It also shows that the higher the mass number, the less efficient the fusion becomes in terms of producing energy (it also takes up less and less of the star's lifetime.

The figure also shows that you can extract energy from the 'other end' – from the heavier elements by decay to lighter elements, although in general this is much more 'milder' process. Nevertheless, the slow decay of these heavier elements is an important process in heating the Earth and other planets (atomic fission as used in power stations and atomic bombs is something different – there a nucleus splits in two and releases a fair amount of energy in one go. If a large number of nuclei can be induced to do the same thing simultaneously then an enormous amount of energy can be released).

Bright blue stars have a Main Sequence lifetime of a few million or tens of millions of years.

Cepheids

Before I talk about Supernovas, can I refer back to Figure 17 – HR diagram showing early T-Tauri starswhich shows the region of the HR diagram inhabited by **Cepheids**[58], implying that these are mostly supergiants. The shaded Cepheid region and the region below it describes a region called the 'instability strip' through which most stars are assumed to pass during their evolution.

---

[58] pronounced 'see-feeds' officially

Cepheids themselves are 'pulsating' stars, with these pulsations occurring on a regular basis. These stars are extremely important stars for reckoning distance – their pulsations are directly related to their intrinsic luminosity.

This property was first discovered by Henrietta Leavitt working at the Harvard College Observatory. It is a feature of the 'land of the free' that until recently women were barred from working in high academic positions. At the end of the 19th/beginning of the 20th century, the Harvard College Observatory was headed by Professor Edward Pickering who had a team of female 'computers' who were hired to do the basic work of calculating. Several of these 'computers' rose to do more responsible work, without actually achieving the official status that would be afforded them today.

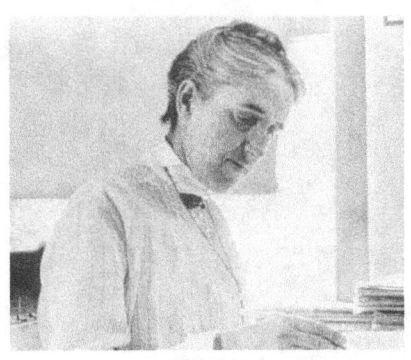

Figure 20 - H. Leavitt

Henrietta Leavitt was one of these. She started to investigate Cepheids in the Small Magellanic Cloud and used the assumption that all the stars were at the same distance away from us (which is indeed valid to a good approximation, although at the time they had no idea of the exact distance of the Small Magellanic Cloud). Leavitt was able to observe there was a direct mathematical relationship between the apparent luminosity and the pulsation period of the Cepheids – the greater the luminosity, the greater the period. Using the assumption just mentioned, the relationship must also be true if apparent luminosity is replaced by inherent luminosity. These results became known in 1912 but since no actual distances were yet known for any Cepheid, then the scheme could not be used immediately to calculate actual distances.

Physically, Cepheids increase and decrease in magnitude over a period of from 1 day to about 100 days. Its radius typically increases by about 10%. Polaris was a well-known Cepheid but very recently it appears to have stopped pulsating.

Classes of variable stars tend to be named after their progenitor, and the original 'Cepheid' was δ-Cephei, whose pulsating nature was discovered by John Goodricke in 1784.

Supernova

Anyway, according to theory a star that is fusing iron in its core will eventually go supernova – in this case a **Type 2 supernova**, its processes being different from the explosive type 1a mentioned previously, but looking similar externally with a massive ejection of the envelope at speeds of about 10,000 km/s.

In reality, the exact sequence of events is still under dispute by scientists but the theory does depend on a rapid collapse of the (approx. Sun-sized) core, so rapid (about 0.1 seconds or less) that the envelope has effectively no time to feel the effects. This core collapse produces an enormous outflow of neutrinos (whose energy accounts for 99% of the energy emitted during a supernova) and if the collapse can be halted by forces exerted by neutrons, the resultant rapid bounce will produce a mighty shock wave.

Both the neutrinos and the shock wave are candidates for ejecting the envelope, but the theory seems uncertain at the moment.

I have mentioned neutrons as being an important factor in halting the collapse. There is an excess of neutrons in the core because an inverse beta reaction[59] has taken place, whereby electrons and protons[60] have combined to form neutrons by means of an 'inverse beta reaction'

$$p + e^- \rightarrow n + v$$

(a proton and an electron combine to form a neutron and an anti-neutrino)

The removal of electrons in this manner will speed up the collapse because they were tending to resist the collapse by virtue of their degenerate state – the same degenerate state that stops a white dwarf from collapsing. The exact manner of the inverse beta reactions is a factor affecting the collapse - its influence on the time of collapse is an important consideration in the 'shock-wave theory'.

You end up with a core composed mostly of neutrons and if these neutrons do halt the collapse (by virtue of neutron degeneracy[61]) then, although the envelope will be ejected, the core will remain – as a **neutron star**. Whereas white dwarfs are 'held up' by electron degeneracy, neutron stars are 'held up' by neutron degeneracy[62].

The star will be about the size of Manhattan, which I always say to students[63] is ironic because Robert Oppenheimer was behind the initial mathematical theories of neutron stars and later led the Manhattan project – the project to produce an atom bomb.[64]

Normally neutrinos are extremely 'unreactive' , but if only 1% of their energy is imparted to the envelope then that would be sufficient to produce the observed effects of envelope ejection (and this transfer of energy *could* be possible under the extreme conditions prevailing).

---

[59] The beta reaction is $n = p + e^- + \bar{v}$ whereby an neutron in an unstable nucleus turns into a proton and emits an electron (the so-called beta particle) and a neutrino
[60] There has been a process of 'photodisintegration' whereby high-energy photons have split the iron nuclei into their constituent protons and neutrons
[61] the analogous quantum-mechanical concept as the previously mentioned 'electron degeneracy'
[62] The maximum mass of a neutron star is 2 solar masses, although this particular 'limit' is not given a name as far as I know
[63] by way of a mnemonic
[64] Oppenheimer had worked with Synder on black holes in 1939. Work with Volkov about the same time resulted in the Oppenheimer-Volkov Limit of a neutron star, its upper limit, currently estimated to be about 2-3 solar masses.

Shock waves produced by aircraft wings typically slow supersonic flow down to subsonic velocities, which is equivalent (when you think about it) to imparting a velocity to the airflow in the direction of travel of the shock wave. Thus a shock wave moving through the envelope will impart a velocity to the envelope in the direction of the shock wave. We are looking at the possibility of a very strong shock wave capable of ejecting the envelope, and we need to take into account that any shock wave will 'weaken' as it imparts energy to the envelope[65]. The shock-wave scenario is sensitive to the speed of the collapse – in general, the faster the collapse, the more successful is this shock-wave model.

If the neutrons are unable to stop the collapse, as might be the case in more massive stars, then we will have a **Black Hole** produced.

Obviously evidence for black holes is less than 100% certain, but there is strong evidence for their existence. Primary areas of research are binary systems where one of the components is 'invisible' but obviously having a gravitational effect on its companion.

Alternatively, a Black Hole could be surrounded by an accretion disk – which due to the strong gravitational field would be emitting a considerable amount of high-energy radiation. The black hole itself can never be seen.

If a black hole had no accretion disk (and is not in a binary system), then it would not be readily observable. The idea that a massive black hole (produced in a different manner to the one just described) lies at the heart of many galaxies, including our own, is widely believed, although without an accretion disk they are more elusive. Their gravitational effect by itself tells us nothing – all other things being equal, the gravitational effect would be identical to that of 'normal' matter of equal mass. Using a thought experiment, if the Sun were able to shrink smoothly to a 2cm diameter black hole, everything in orbit about the Sun would keep on moving exactly as before.

Supernova Detection

The number of supernovas actually seen visually in our own galaxy has been very low. The last one was in 1604 (by Kepler)[66], despite the predictions (backed by observations of other galaxies) that they occur about every thirty years or so on average.

The problem is dust, which reduces our radius of vision to about 1kpc[67].

---

[65] The most common everyday type of shock wave is not due to aircraft, but is in fact thunder. It is no surprise that a shock wave will weaken in the way described above, but this is obviously something you will have experienced in real-life (thunder dissipates completely after about 20 seconds)

[66] and before telescopes were invented!

[67] At 1kpc distance, it is impossible to see a solar-type star.

Apart from 1604, only about half a dozen supernovas have been observed visually in the Milky Way[68], including one in 1572 which was studied by Tycho Brahe and one in 1054 which produced M1, the Crab Nebula. The latter was recorded in Chinese records but seemingly not in the West at all.

However, a 'nearby' supernova did occur in 1987 in the Large Magellanic Cloud[69] – this is named 1987A, i.e. the first supernova to be spotted in 1987. This was first noticed by a professional astronomer in Chile, although another astronomer in Australia later spotted it on photographs he had taken 19 hours earlier.

Three hours before this 'Australian sighting' four detectors on Earth recorded a total of about twenty neutrinos detections, probably from the supernova. These neutrinos had travelled to us at the speed of light, effectively from the time of collapse. Supernova brightening and emission of light occurs later.

Paradoxically, of these detectors the two most 'successful' had been built to detect something else completely – the decay rate of the proton. But nowadays, there are even more sophisticated devices in position, up and ready to record neutrino emissions from any nearby supernova (the Supernova Early Warning System). Any such fore-warning from a neutrino detection would allow visual observation right from the beginning (which has never happened before).

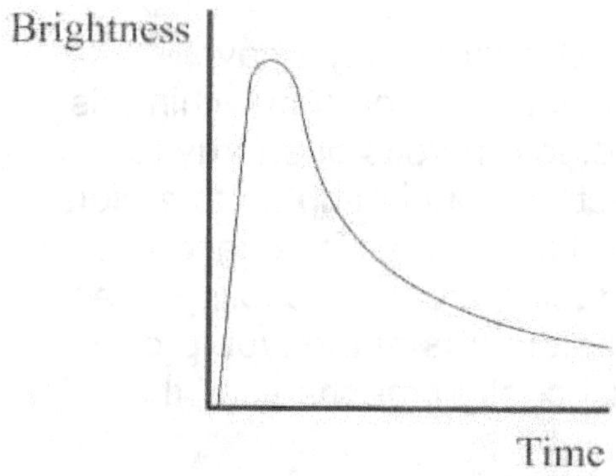

Figure 21 – typical light curveshows a typical light curve for a type II supernova.

Figure 21 – typical light curve for a Type II supernova

Supernovas in other galaxies are however often detected – so much so that this an area where amateurs play an important role. Not so long ago, this fundamentally required extreme patience – to observe a large number of galaxies regularly and remember their appearance so that anything unusual can be detected straightaway. Nowadays, computer technology has helped considerably.

Nomenclature has responded accordingly. Traditionally, the first 26 supernovas of the year were designated with a capital letter from A to Z, and until 1988 this system was usually sufficient. However since then each year has required the use of an extension (which did already exist but was rarely used) - pairs of lower-case letters are used: *aa, ab,... ba, bb, ..* etc. etc. As

---

[68] However, as of 2000, about 180 supernova remnants were known in our galaxy, most of them only detectable via radio. Five are visible either to the naked eye or photography – the Crab Nebula, Veil (or Cirrus) Nebula in Cygnus, IC443 in Gemini, S147 in Taurus and the Vela (part of a much larger nebular complex called the Gum Nebula). A sixth can be picked up by CCDs or long-exposure photography – Kepler's supernova of 1604 in Ophiuchus.
[69] The Small Magellanic Cloud and Large Magellanic Cloud are satellite galaxies of the Milky Way

an example, 572 supernovas were discovered in 2007, 261 in 2008, 390 in 2009; 231 in 2013.

Features of Supernovas

It has been mentioned that only hydrogen and helium were produced in the early Universe, soon after the Big Bang. All other elements stem from supernovas. In addition to the elements up to and including iron which they have already produced, in the short time available during the destructive event itself they produce elements heavier than iron and 'spew out' almost everything into space, or what is called the **Inter-Stellar Medium** . In this sense we ourselves are indeed 'stardust', because our bodies require these heavier elements created by supernovas.

Younger stars will also contain *traces* of these heavier elements and this has produced a division into two star classes – paradoxically the younger stars are classed as **Population I stars**, whereas **Population II stars** are the older stars formed only from hydrogen and helium. There is no clear-cut boundary separating the two populations, but you will see this classification used quite often.

These additional elements are also relevant to star formation. I have mentioned star formation through cloud contraction If a collision within this collapsing cloud is capable of imparting 13.6 electron volts of energy to the orbiting electron of an hydrogen atom, this electron could jump up to a higher level. If it was later to drop down to a lower level it would emit radiation which could then maybe escape from the cloud, producing a certain cooling effect. The point is that trace molecules, formed from elements of the type produced in a supernova, will be much more efficient this 'cooling mechanism' than hydrogen.

The effect is this – in the early Universe with clouds of hydrogen and helium only, collapsing clouds found it 'difficult' to radiate energy and stars tended to be bigger. In the later Universe, thanks to these trace substances, a cloud will radiate more easily and thus tend to become more unstable and fragment into a larger number of smaller stars. This theory is potentially testable – the earlier stars will tend to produce more supernovas (being larger on the whole) and these supernovas will seed the Universe with elements at a particular rate which would be faster than today. The abundance of elements in the Universe needs to match this estimated output from earlier stars.

Now I have mentioned two types of supernovas, I could just list a few features

Type I	Type II
absence of hydrogen in spectrum (consistent with a white dwarf which will have thrown off its outer layer during planetary nebula phase)	hydrogen lines in spectrum
favored cause of Type 1a – explosion of a white dwarf in a binary system, accruing matter from its companion sending it over its Chandrasekhar Limit	'death' of a massive star
On average, brighter than Type II	
possible correlation between Type I supernovas and older elliptical galaxies	many are found in arms of spiral galaxies, which are star forming regions of the galaxy

Neutron Stars and Pulsars

Neutron stars escaped from the realm of pure theoretical studies in 1967. Jocelyn Bell, a research student at Cambridge, detected regular repeated emissions from space, of the type that might have been expected more from extra-terrestrials than from astronomical objects – the emission from the original pulsar occurred regularly every 1.33 seconds. These objects, whatever they might be, were christened pulsars by astronomers

Very shortly afterwards they became identified with neutron stars, which is now accepted theory. Due to the manner of its creation a neutron star would be expected to have a high rate of rotation (compare the ice-skater who speeds up as her/his 'diameter' becomes smaller). Its size is small enough to rotate at the observed rate without ripping itself apart, so if it was emitting radiation from a point on its surface, in a manner similar to a lighthouse, then this regular pulsing would be detected by us if the Earth happened to lie in the path of the 'beam'.

The first pulsar to be seen in the visual spectrum was the pulsar in the Crab Nebula[70], in 1968 – this has a period of 0.0335 seconds. This particular object played an important part in establishing the theory – the energy given out by the pulsar proved to be consistent with the energy required to power the supernova remnant since 1054.

Pulsars have been detected with a period as low as 0.001558s and as high as 4.3s. In 1974 the Binary Pulsar was discovered[71]. Since then a couple of other binary pulsars have been discovered

Black Holes

Popular treatments of black holes mostly deal with what are called **Schwarzschild** black holes. These holes are static, and once anything

---

[70] the remnant of a supernova from 1054 (and designated M1 in Messier's list)
[71] And both supervisor and research student received the Nobel Prize, in contrast to Jocelyn Bell, who was not awarded the Nobel Prize while her supervisor Antony Hewish was

crosses the **event horizon**, they cannot communicate with anything outside the event horizon and must also hit the point singularity at the center of the region. Real-life black holes are likely to be rapidly spinning and will be described by another model due to **Kerr.** In a Kerr black hole the singularity is a ring and any body crossing the event horizon is not doomed to hit it.

Particle Physics

Supernovas are also of interest to particle physicists. The theory at the moment have all particles arranged in two groups, the quarks and leptons – each of which have at least three generations, as follows

Quarks	Leptons
top, bottom	т-particle (tau particle), tau neutrino
strange, charm	muon, muon neutrino
up, down	electron, neutrino[72]

The bottom generation corresponds to 'everyday' matter – they occupy the lowest energy level and predominate at our current energy levels. In the past, nearer the time of the Big Bang, the higher energy generations would expect to be more numerous[73], and it is conditions near the time of the Big Bang that particle accelerators are effectively replicating. The more powerful they become, the 'closer' they come to the conditions of the Big Bang itself.

Now the question is this - are there more than three generations? Traditionally, particle physicists can try and tackle this question by building bigger and bigger particle accelerators, but supernovas offer their own way to 'solve' the problem, as follows

Practically all the energy given out in a Type II supernova 'explosion' is in the form of neutrinos, and this energy is believed to be shared out equally among all types of neutrinos. So, simply put, if we on Earth measure the flux of one type of neutrino and then divide this into the total energy calculated for the entire 'explosion', the answer should be the total number of generations.[74]

Solar neutrinos There is a long-standing problem with solar neutrinos. These escape directly from the Sun and any we detect were only produced about 8 minutes earlier. But the flux we are detecting is only one-third of what we expect theoretically. But if there were three flavors and all neutrinos 'oscillated' into different types of neutrinos, then this could be a solution.

---

[72] Sometimes referred to a 'electron neutrinos'

[73] At least one of the particles mentioned, the muon, can be detected fairly regularly as a result of cosmic ray collisions in the upper atmosphere. They only have a short life as they cascade towards Earth.

[74] To be more precise the total supernova energy will be divided out equally among all neutrinos and anti-neutrinos. So, for example, if the flux of electron neutrinos was found to be one-sixth of the total supernova energy, this would be an indication of three generations

Neutrino Mass

In the past, neutrinos have been assumed to be massless, and therefore they have to travel at the speed of light. New theories consider whether they have a <u>minute</u> mass – if so, they cannot travel at the speed of light. Given that the electron neutrino, muon neutrino and tau neutrino would then have progressively more mass, then they would arrive at detectors in three separate waves. This mass has enormous consequences for cosmology..

Endnote

This description of stellar evolution has mostly been concerned with single stars. About half of stars are binary and this binary nature is likely to modify the evolution of the component stars

1987A

Research is still continuing on this object, although no pulsar has yet been detected. As the matter slams into the interstellar medium, it can produce visual 'rings. Using knowledge of the speed of light, the dimension of this ring can be calculated – and by comparison with the size that we see, we can calculate the distance to the supernova.

Chapter 4   Messier Objects, Galaxies

Messier Objects

The more obvious non-stellar and non-planetary objects in the (Northern) sky usually have a 'Messier' designation. Messier was a French cometary observer of the late 18th century who compiled a list of non-cometary objects so as to avoid confusion in his work[75]. Nowadays the Messier list contains 110 objects[76].

Messier himself spent 50 years searching for comets and found about 20. There is such a thing as a Messier Marathon which involves viewing all 110 objects in one night. This is impossible to do from Britain, but can be done from the likes of Portugal.

Messier objects belong to several categories

- Spiral Galaxies, similar to our own Milky Way. M31 the Andromeda galaxy is a famous example, a naked-eye object. The spiral nature of these objects was only discovered in 1845-48 by Parsons (Lord Rosse), using his fixed telescope in Ireland (still the largest telescope in the world at the dawn of the 20th century).

- Elliptical Galaxies, whose shape is self-explanatory from their name. In general, they contain almost exclusively older stars.

- Galactic (or Open) Clusters clusters of stars found mostly within the galactic disk. I have already mentioned the possibility that a collapsing cloud will fragment into several stars. The resultant clusters will disperse over time for various reasons, e.g. as individual stars 'evaporate'[77] but still-existing clusters would be expected to be found in the galactic disk where the star-forming regions are. Famous examples are the Pleiades (M45), and Praesepe (M44) sometimes referred to as 'The Beehive'.

- Globular Clusters clusters composed of many older stars, typically 100,000 to 1 million, which are bunched more tightly than open clusters. These clusters tend to be found in the halo of the galaxy, not in the disk. There are no naked-eye globulars in the North (except possibly M13 under exceptional circumstances). However Omega Centauri and 47 Tucanae are both naked-eye objects in the South.

---

[75] In 1758 he was searching for Halley's Comet on its predicted return, calculated to be in Taurus. It was during this period that he first saw the Crab Nebula which he originally thought might be the comet (and which became M1 in his list).
[76] It would not be fully accurate however to state that these 110 objects are exactly the 110 brightest objects in the northern sky
[77] Individual stars will receive enough energy to escape in a manner similar to individual water molecules receiving enough energy to escape from a liquid

Such are the uncertainties in the age of the Universe that sometimes globular clusters are found to be older than the Universe itself!

- <u>Planetary Nebulas</u> which have already been mentioned in Chapter 2. The best known is M57, the Ring Nebula in Lyra, (although the brightest is NGC7293 in Aquarius, which is not in Messier's list)

- <u>Supernova Remnant</u> there is only one of these but it is the first on the list, M1 the Crab Nebula (its name derives from its appearance as seen from Rosse's telescope). This is the remnant of a Type 2 supernova of 1056. This was well-documented at the time in the Orient but not in the West at all. Prevailing dogma at the time appears to have followed Aristotle's idea that any changes in the heavens of this type could not exist, and the population seems to have responded in a manner still recognizable today. There are echoes of George Orwell's 1984: 'the party instructs you to disregard the evidence of your eyes and ears'[78].

- <u>Emission Nebulas</u>[79] These clouds are emitting radiation on their own account by virtue of its gas being heated by stars embedded within or near the nebula. The best-known example is M42, the Orion Nebula, which has already been briefly mentioned (to be exact, planetary nebulas are also a form of emission nebula).

The New General Catalog (NGC) is a larger collection of non-stellar objects, based on the work of John Herschel[80] but expanded by J. Dreyer to encompass 7840 objects. Recently. there have been attempts to extend the challenge for amateurs who have completed a Messier Marathon by producing a 'Herschel 400'[81]. All Messier objects are also included in the NGC – for example, the Crab Nebula is NGC 1952 and M31 is NGC224[82].

Galaxies

Observing

Finding galaxies is not as easy as some books make it out to be. One piece of advice I have come across is – if this is your first try at seeing galaxies, start out by removing from your mind any images of galaxies you may have seen in photographs. What you will see is faint light from millions of stars that has been travelling across space for about 50 million years before it hits your retina. Further, the magnitude of galaxies can be misleading. An 8th

---

[78] or alternatively, the 'Emperor's New Clothes' by Hans Christian Anderson
[79] there are other types of nebula such as **reflection nebulas** where dust reflects light from elsewhere. **Dark nebulas** are dense clouds capable of obscuring background objects – examples are the Horsehead Nebula in Orion (which is at a temperature of 10K), the Coalsack in the Southern sky, dark lanes in the Milky Way within Cygnus, Aquila and Sagittarius
[80] By 1864, he had catalogued 2478 objects
[81] other similar lists have also been put forward
[82] It might be interesting to note here that the Double Cluster in Perseus (which are absent from Messier's list) are classified as NGC 869 and NGC 884 . Perversely they had previously been given star names as h- and χ-Persei

magnitude galaxy is not like an 8th magnitude star - rather the light in an 8th magnitude galaxy is spread out over a region of sky.

History

At the beginning of the 20th century, prevailing opinion appears to have been that objects like the M31 Andromeda 'Nebula' (as it was then known) were just nebulas within our own galaxy. There were adherents to the idea of 'island Universes' and this lead to what is known as the 'Great Debate' of 1920, principally between Curtis on the 'island Universe' side and Harlow Shapley on 'the Milky Way is the only galaxy' side. The meeting was inconclusive.

The issue was resolved in 1924 by Hubble who was able to detect a Cepheid in M31, and using the recent luminosity / pulsation period rule was able to calculate that M31 was about a million light years away. It was definitely a galaxy in its own right[83].

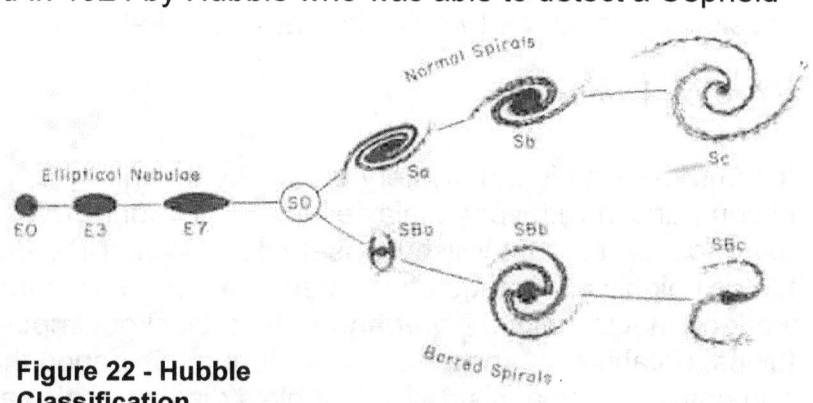

Figure 22 - Hubble Classification

Hubble was able to follow this up by producing a classification for galaxies. The main division is into spiral and elliptical galaxies, with the rest being classed as 'irregular'. The adjacent diagram is a bit misleading in that it implies some sort of evolution whereas if there is a connection between the different types of galaxies, there is currently no general agreement as to what it is[84]

Spirals are further divided into 'normal' spirals and barred spirals, so-called for reasons obvious from the diagram. An Sa (or SBa) will have large bulges and tightly wound spirals. You can see that going through b to c, their bulges decrease in size and their arms become less tightly wound.

An S0 galaxy is loosely considered as a 'spiral without any spiral arms'.

The ellipticals are classified from E0 to E7 depending on their degree of 'flattening'[85]. I should stress that this degree of flattening is not necessarily an inherent characteristic of the galaxy – it purely relates to the shape as we see

---

[83] The fact that it was not recognized as a galaxy earlier also reveals how little was known about what our own galaxy looks like.

[84] To quote just one example, some theorists attempt to show an elliptical forming from the collision of two spirals but this is all work 'under progress'

[85] The ellipticity é is given by e = (a-b)/a, where a and b are the semi-major and semi-minor axes respectively.   The number in E0 –E7 is ten times the relevant ellipticity

it, which can obviously depend on the aspect from which we are viewing a particular galaxy.

Ellipticals	Spirals
stars moving in random directions	mostly moving in near-circular orbits (in the disk anyway)
little gas	plenty of gas
plenty of older Population II stars	new Population 1 stars in disk as well as older Population II stars in the bulge and globular clusters which tend to orbit in the halo

60% of galaxies are ellipticals and 30% spirals. Of the spirals 60% are barred. Despite the assumption up to now that we live in a normal spiral, there is possible evidence that we live in a barred spiral.

More on Distance

It came to be a bit of a mystery as to why the Milky Way appeared to so large in comparison with spiral galaxies like M31. During World War 2, a German astronomer at Mount Wilson Observatory, Walter Baade, suddenly found himself highly advantaged by a) being excluded from the obligation to do any work connected with the war and b) by a blackout imposed on neighboring towns, notably Los Angeles. It was during this period that he observed M31 and came across a mistake in Hubble's distance calculation. His observations were able to detect more obscure Cepheids and he realised that there were different classes of Cepheids – Hubble had used the wrong class.
in 1954 Baade was able to formally announce that the distance to M31 was two million light years - double Hubble's figure. Our galaxy was no longer anomalously large, and what's more this new scale meant that in fact the entire Universe was twice as large as had previously been thought.

Since then the scale has been modified again, M31 is now considered to be 2.9 million light years distant.

Nature of Spirals

As shown in Figure 23 - A Spiral Galaxy, and has already been mentioned, a spiral will possess a nuclear bulge, which is composed mostly of older Population II stars. It has a disk and when seen in 'plan view' a noticeable feature of the disk are the spiral arms. The whole thing is surrounded by a less visible 'halo' formed from hot intercloud material[86].

The spiral arms do not actually contain a larger number of stars than elsewhere in the disk – what they do contain is a larger number of bright stars, and HII regions[87], because they are star-forming regions. This is consistent

---

[86] 'Hot' infers that the material is ionic, 'warm' implies it is atomic, and 'cold' implies it is molecular
[87] HII regions are star-forming nebular regions

with the theory that I have already mentioned – brighter stars tend to have shorter lives and will obviously therefore make star-forming regions brighter than regions in the rest of the disk containing older stars.

This leads me to mention an important problem. The spirals are not actually rotating as a single 'material' entity. If they were, they would soon 'wind-up'. Think of the Solar System – there the planets are **not** rotating in a manner similar to a wheel (or a rigid body), the inner planets are rotating around the Sun in about one or two Earth years whereas Neptune requires 165 Earth years (this type of motion is called differential rotation). If the spiral arms were rotating in a like manner, they would soon lose their shape. Spiral arms can be seen to be rotating in manner similar to (very slow) Catherine wheels but the theories for explaining this are still 'in the air'. Quite definitely, the rotation of the spiral arms and the rotation of the disk itself are two separate things. The rotation of the latter is 'straightforward' in terms of gravitational theory. Objects in the disk rotate in a manner similar to the Solar System but modified by virtue of the fact that there is no *dominating* central mass, like there is in the Solar System. The rotation is nevertheless differential rotation[88].

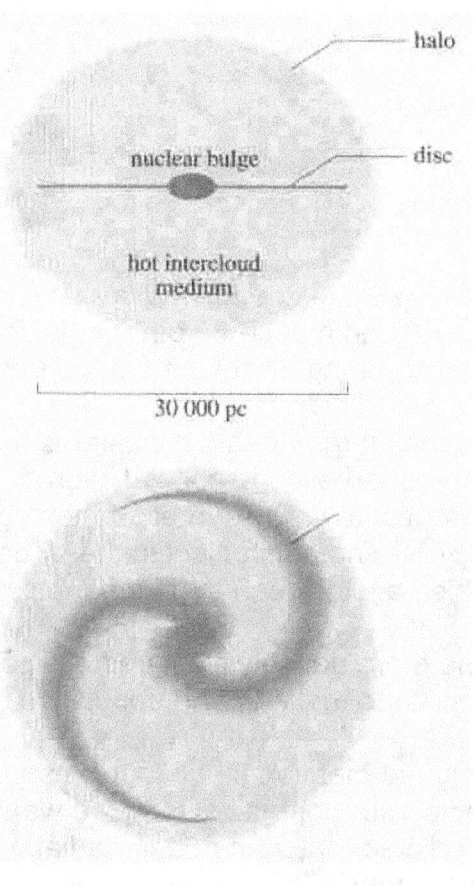

Figure 23 - A Spiral Galaxy

Note that in the diagram, distances are given in parsecs (pc). These are easier to remember than light years. For the Milky Way (which is a quite typical spiral)

Diameter of galaxy   40 kpc[89]
Diameter of Disk :   30 kpc
Diameter of Bulge: 6 kpc
Width of Disk;   1 kpc

---

[88] So there is no single rotation rate for the Galaxy as is sometimes implied
[89] kpc – kiloparsec, i.e.. one thousand parsecs

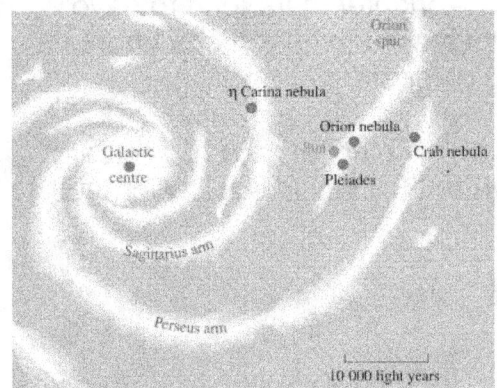

Figure 24 - Our position in the Milky Way

The Sun lies 20 kpc from the galactic center, in the Orion Arm

Dark matter in the Galaxy (matter which cannot be seen but is exercising a gravitational effect) might amount to between 4-60 times the visual matter

Larger-scale structure.
Galaxies themselves form themselves into clusters of galaxies. A cluster of less than 50 members is called a group and we are a part of the Local Group which is dominated by the Milky Way and M31 and which also includes M32, M33[90], M110, and maybe 30 other galaxies. It is instructive to realize that galaxies are still being discovered in the Local Group, e.g. the Antlia galaxy in 1997.

A much larger 'nearby' cluster is the Virgo Cluster with 2500 galaxies and if you observe in the constellation of Virgo and its environs, you have the potential to view several of these galaxies fairly easily (see Figure 13 - Virgo). M87 is a notable object in the Virgo Cluster, a giant elliptical galaxy, from which a jet is seen to be emanating.

The more densely clustered are galaxies, the more that elliptical galaxies appear to make up the population.

I should mention that a lot of research into our own galaxy has been done by radio astronomers. During the war, Van der Hulst in the Netherlands had predicted that 21 cm radio radiation could be detected from galactic hydrogen[91]

Active Galaxies

Various 'unusual' forms of galaxies began to be observed, although they still seem to sit within the Hubble Classification - some are elliptical and some spiral

Quasars are probably the most well-known of this class. They received their name originally because they looked like unusual stars and had been detected by radio telescopes (quasi-stellar radio stars). An object called 3C-273 was a known to be a strong radio emitter[92] and in 1962, the Parkes radio telescope in Australia was able to identify it with a visible object. It was

---

[90] M33 is probably a satellite of M31, and probably many of the Local Group could also be satellites of other galaxies

[91] This is due to a flip in the spin from plus half to minus half, or vice-versa, an extremely rare event with a probability of occurring in an atom every 10 million years. But there is so much atomic hydrogen in the Universe that this radiation is detectable. Note also that the density of hydrogen atoms is so low that such conditions would be classed a perfect vacuum if reproduced in a laboratory (which it can't be).

[92] It gets its name by being contained in the 3rd Cambridge Catalog of Radio Stars of 1962

recognized by Maarten Schmidt in 1963 that they were actually highly red-shifted objects and were gradually recognized as highly-compact and extremely energetic objects and then eventually that they were extraordinarily energetic galactic cores.

Not surprisingly, these energetic cores are assumed to derive their energy from black holes and are assumed to be very distant, and therefore symptomatic of a class common in earlier times. Analogous to what I have said previously about stars, these objects could be active purely because they are surrounded by matter energized by the gravitational attraction of the black hole. They could 'settle down' when this matter is 'used-up' and therefore these objects could be an earlier form of the 'more placid' nearby galaxies – the black hole will still be there but its influence will be less noticeable

The main classes of **active galaxies** are

Quasars   Although having a name that implies they are radio-loud objects, only about 10% are indeed radio-loud. It is only with the likes of the Hubble Space Telescope that the host galaxy can be seen, and these imply that quasars are contained within elliptical galaxies

Radio Galaxies   Characterized by two radio lobes either side of the active core and in many cases a jet, or jets, of radio-emitting material emerging from a compact central source at about 1000km/s. It can radiate up to 1000 times the energy emitted by an entire galaxy. The first radio galaxy to be discovered was Cygnus A . In 1951 this was shown to coincide with an unusual elliptical galaxy, and all radio galaxies have been found to be within ellipticals.

Seyfert Galaxies   First identified in 1943 by Carl Seyfert. Less luminous than quasars but displaying similar characteristics including short-term variability. A Type 1 has broad spectrum lines, whereas a Type II has narrow lines. M77 is a Seyfert Galaxy

BLac (also referred to sometimes as 'blazars'). These were originally classified as variable stars in 1941. In 1968 a connection was made with powerful radio sources, and were found to have a featureless spectrum. From 1974, a connection was made with galaxies. BLacs were investigated by the Compton probe in 1991 and it was noted that gamma ray emission is as powerful as any other wavelength.

2% of galaxies are active.

The theory you will see nowadays is that all these objects are actually one and the same type of object viewed from different angles. In this model, a torus of dust circles the active center. Then a Type 1 Seyfert could be due to fact that we can see the very active regions close to the active center, whereas a Type 2 will result from a viewpoint where the torus cuts off the view of the most active regions, allowing only a 'view' of the less energetic regions farther out.

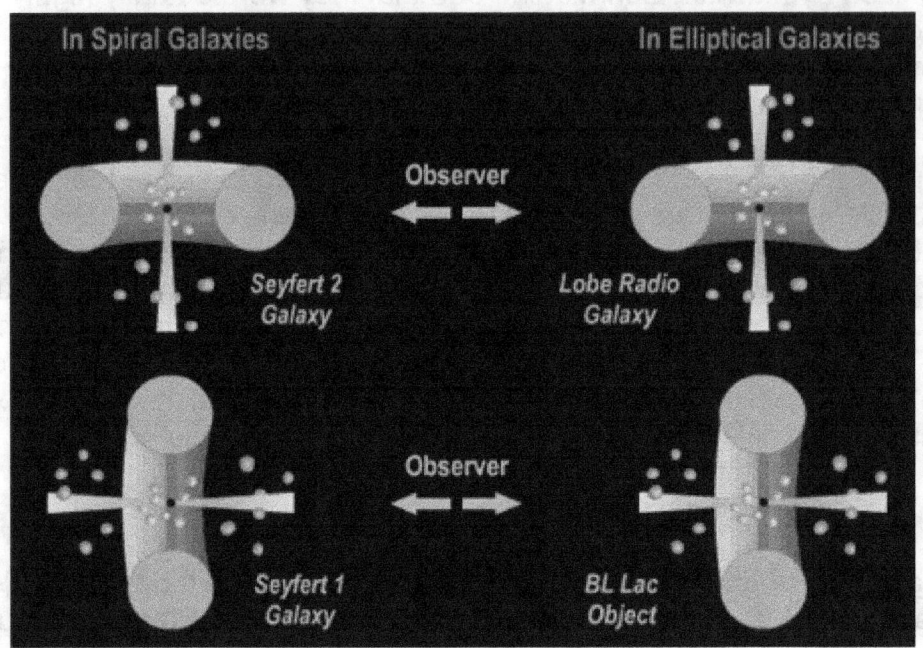

Starburst Galaxies

These are spiral galaxies that are emitting far more infra-red than an 'ordinary' galaxy. The common theory is that two galaxies have collided and stimulated star production. M82 in Ursa Major is a well-known example.

Chapter 5   The Sun

As already alluded to, the Sun is a lower mass Main Sequence star (i.e. hydrogen burning) with a surface temperature of 5800K and a core temperature of 15 million degrees. Its light takes approx. 8 minutes to reach us.

It is composed of roughly 75% hydrogen and 25% helium (by mass)[93], but with about 1.5% of other elements. Four million tonnes of mass is being converted to energy per second within the core. The Sun is about halfway through its estimated life of 10 billion years.

Surface

The 'surface' that we see when we 'look' at the Sun is the **photosphere** – it is actually a layer about 500km thick[94]. At the bottom of the photosphere the temperature is about 9000K and at the top about 4500K. The mixture of light[95] that we see from different depths gives the effective temperature of 5800K which we usually quote as being the Sun's surface temperature. There are two important layers above the photosphere – the thin **chromosphere**[96] and the more extended **corona**, whose appearance varies noticeably during the eleven-year solar cycle. These latter two regions are only visible to laypersons during a solar eclipse – at other times they are swamped by light from the photosphere.

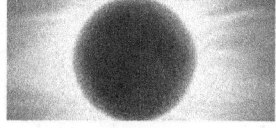

**Figure 25 -
Extended Corona**

Hydrogen alpha filters are quite commonly used with conventional telescopes to allow viewing of the chromosphere. The dominant H alpha spectral line produces a pink/reddish color in the chromosphere, thus giving the layer its name. It was named as such by Joseph Norman Lockyer, who also gave helium its name after it was discovered in the Sun.

During the 11 year solar cycle (which is actually a 22 year cycle because the polarity alternates between each '11 year cycle') the corona becomes most extended at the so-called *Solar Maximum* and less-extended at *Solar Minimum*[97]. At Solar Minimum, it might just consist of a couple of streamers either side of the equatorial regions.

The temperature of the corona is around 2 million degrees[98] - but although the corona might be at a very high temperature, it is very tenuous. This situation

---

[93] but a ratio of 12:1 in terms of particles

[94] We also 'see' light from different 'depths' when we look at things like clouds

[95] and Infra Red

[96] about 5,000 km thick, at a temperature of 4,000 to 10,000 degrees, cf. the Sun's diameter of 1.4 million km

[97] Strictly speaking. the 'cycle' could vary between 8 to 15 years

[98] In science, temperature is a measure of the average velocity of the atoms/molecules making up the substance. So the high temperature of the corona indicates that the particles in the corona are moving very fast but because the corona is relatively thin, but the actual amount of heat there is not as high as the layperson might think. Think of whether you would

is a major mystery in astronomy because the corona cannot be heated to this high temperature by a straightforward transfer of heat from the lower temperature photosphere (this is a statement of the Second Law of Thermodynamics). Observations in 1997/8 by SOHO[99] seems to confirm theories that the heating is caused by magnetic fields.

Features

Many features can be noted on the Sun which are associated with emerging magnetic fields, for example sunspots.

The discovery of the existence of sunspots is usually attributed to Galileo, although again the Chinese appear to have been noting them for quite a few centuries beforehand.

These spots are not actually dark, they appear dark to the human eye because of a contrast effect with the hotter 'surface (only hotter by about 1000 degrees or so!)[100]. They come in pairs of opposite polarity[101] - in one hemisphere, all pairs will be lead by a spot of the same polarity while in the other hemisphere all the pairs will be lead by spots of the opposite polarity. After an 11 year cycle, all these polarities will reverse – effectively the Sun itself is reversing its polarity.

At the start of a cycle, sunspots appear at about 45° latitude. As the cycle progresses, they move towards the equator and become numerous, reaching 30° or so at maximum. They then decrease in number but continue moving down to about 7°, before petering out and re-appearing again at 45°.

Sunspots rotate by virtue of the Sun itself rotating, but this rotation varies with latitude - from 25 days at the equator to maybe 36 at the poles[102].

The best way to observe sunspots is to use a telescope to project the Sun on to a card etc., but keep eager spectators away from the telescope. Sunspots vary in size but no more than 1% of the Sun's face will be covered in spots.

---

put your hand in water at 100 degrees (not recommended) or whether you would put your hand in an oven at 100 degrees (safe). You might get a perfect boiled egg by boiling it for 3 minutes but placing an egg in an oven at 100 degrees for 3 minutes will prove to be disappointing. Other examples of hot objects containing little heat : sparklers (fireworks), sparks from a circular saw

[99] Solar Heliospheric Observatory, satellite administered by ESA (European Space Agency) & NASA

[100] If you could just see the sunspots and not the rest of the Sun, they would appear brighter than any stars as viewed from Earth,

[101] By polarity, we mean the North or South magnetic polarity. This is different from talking about the spots actually being in the North or South hemisphere

[102] Due to the Earth's motion along its orbit, a sunspot rotating every 25 days will actually appear to rotate every 27 days – the time taken to cross the near face of the Sun will be half this (if it exists for that long)

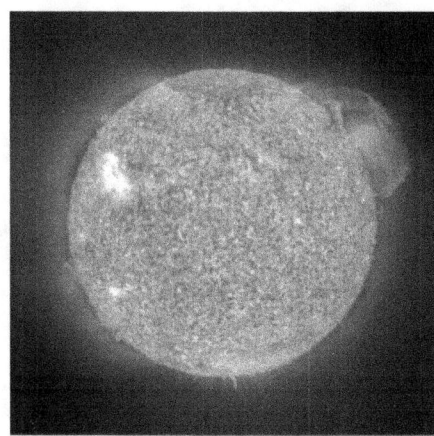

Figure 26 - Sun displaying a prominence

Prominences are visible as arcs when they appear on the limb of the Sun but when viewed on the face of the Sun from a more 'overhead' angle they are called filaments and are dark. **Plages** are bright regions of the chromosphere noticeable around sunspots. Much more energetic than all these are **solar flares**, a dramatic release of energy above sunspot groups, liberating electromagnetic radiation across all frequencies as well as fast-moving charged particles (sometimes near the speed of light), affecting Earth's ionosphere and thereby radio communication. Flares vary with the solar cycle. On Earth we are protected from the dangerous effects of these flares and the solar wind (high-energy particles emitted by the Sun) by the magnetosphere.

The solar wind can also produce auroras, as charged particles oscillate near the Earth's North and South magnetic fields.

Neutrino 'Problem'

Yet another 'famous' mystery posed by the Sun is the 'neutrino problem' (already touched upon). Neutrinos are extremely elusive particles which hardly ever 'react' (about 4 times 10^{14} neutrinos pass through the human body every second, without reacting at all). They stream out from the Sun's core and potentially offer information about the core itself, from 8 minutes previously. This is in stark contrast to the energy that we detect from the surface (light, infra-red etc.) which has typically taken a million years to travel from the core to the Sun's surface[103].

Predictions about neutrino emissions can be derived from nuclear theory. For many years, the actual detection of neutrinos has indicated a deficiency – the number being detected was only about a third of that predicted by theory[104].

More modern and accurate detectors are still showing a definite deficit, although there is support for the theory that the electron neutrinos emitted from the core might change (oscillate) into other forms of neutrino (μ-neutrinos, τ-neutrinos) on the way to Earth, thus accounting for the deficit. This latter (exotic) process would require that neutrinos have a mass.

---

[103] The energy takes this long, but it has changed its 'characteristics' over time. If we picture the energy leaving the core as X-ray photons, then due to absorption and emission as photons 'struggle' to 'find their way out', an X-ray photon will 'have become' a larger number of lower energy photons by the time the energy exits the surface.
[104] Obviously the actual number of detections itself is minute, but these detections allow estimates of total emissions.

Observatories

Modern solar observatories tend to be situated near water. This reduces the deleterious effects of ground heating (although this feature has only really been 'discovered' recently). For example, the Big Bear Observatory is situated in the middle of the Big Bear Lake, at a height of over 2000 meters. The likes of Mauna Kea and Las Palmas have the surrounding ocean as a handy stretch of water.

Helium

Helium was first discovered in the Sun, in 1868 – it was only discovered on Earth in 1895. It is actually very rare on Earth.

Eclipses

The Moon's orbit is tilted about 5° to the ecliptic – if it was not inclined there would be a solar eclipse every new moon and a lunar eclipse every full moon.

As it is, for there to be a solar eclipse, it is necessary that both the Sun and the Moon be at a node (i.e. at an intersection of the ecliptic and the Moon's orbit). By consideration of our 'Celestial Sphere' model, this can only happen twice a year.

But it is not necessary that both the Sun and Moon be exactly at the node – because of their finite size, the Moon can 'partially eclipse' the Sun at a certain distance either side of the node. Because the size of the Sun and Moon in the sky can vary[105] this 'danger zone' (when the Sun can be partially eclipsed by the Moon) can also vary but encompasses a minimum of 30.7°, which will take the Sun about 30 days to cross. Because the synodical month is 29.5 days, you can appreciate it when I say that while the Sun is in the danger zone, it will be eclipsed by the Moon at least once (other factors come into play rather than just the synodical month, but with the small adjustments required the statement just given is still valid).

So you will have at least one solar eclipse every time the Sun crosses the node – and since it crosses a node every six months (because the Moon's orbit and the ecliptic cross at two places), there will be a minimum of two solar eclipses a year. You can apply an analogous logic to lunar eclipses (although with a different size of 'danger zone'), so likewise there must be a minimum of two lunar eclipses each year.

When the size of the danger zone increases there is a larger probability that the Sun will be hit eclipsed by the Moon as it crosses the zone. The maximum size of the danger zone is 37.04° (equivalent to 37.5 days), so we can declare that the maximum number of solar eclipses at each node will be two, and likewise the maximum number of solar eclipses every year is four. By applying the same argument, the maximum number of lunar eclipses is four per year.

---

[105] because of elliptical orbits

Nevertheless, the maximum number of eclipses combined (both solar and lunar) per year is seven. *Note that I am talking about all eclipses here, not just total eclipses*

It has been known by the Chaldeans in Babylon since around 400BC that eclipses occur on a cycle of 6,585 days (18 years 11 days)[106]. This is called the **Saros Cycle.** After every 18 years 11 days[107] (6 585 days), the sequence of eclipses starts repeating itself.

Each individual eclipse is considered to be a member of a designated **Saros Series**, whereby the same eclipse is assumed to repeat itself after each 18 years 11 days, if I can put it like that. Having said that, there are differences on each 'reappearance'. It will re-occur with a difference in longitude of 115° and also a change in latitude.

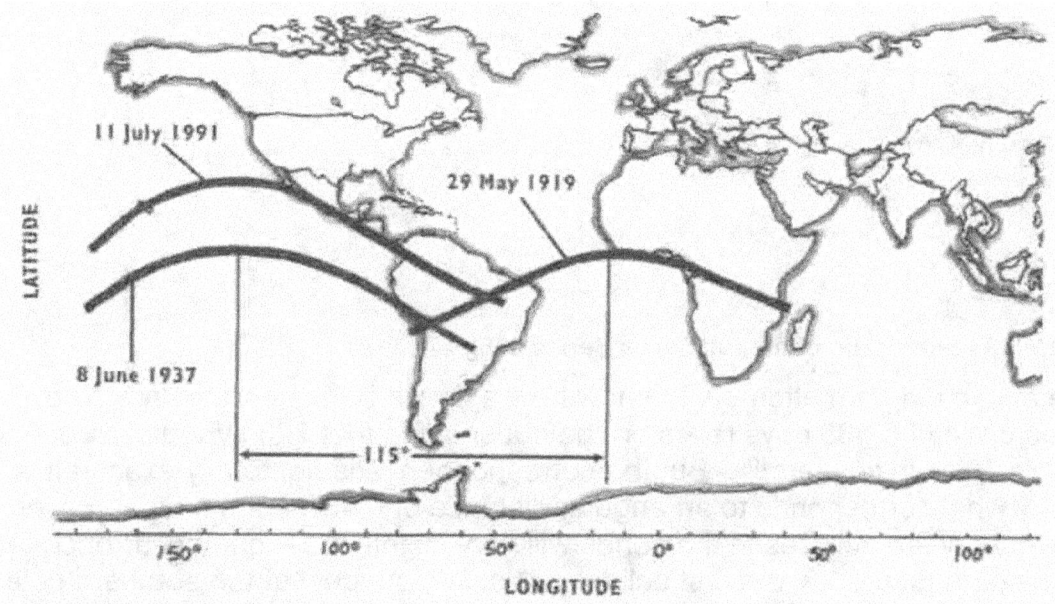

Figure 27 - Saros Series 136

The above figure displays these features graphically for Series 136. A full Saros cycle after the 1919 eclipse it repeats itself 115° further west (note though that there is some overlapping between 1919 and 1937). After a 'triple saros' it 'reappeared' in 1991 at almost the same longitude again, but further North. Eclipses that are moving northwards are given even-numbered Saros Series numbers. Incidentally the 1919 eclipse shown here was the one that was observed by Eddington, on the African island of Principe. and was claimed (in hyperbole published by the press) to have produced evidence for Einstein's Theory of Relativity. In reality, Eddington only produced one picture

---

[106] The book 'Stonehenge Decoded' (Gerald Hawkins) from 1963 implies that Stonehenge could have been used to predict eclipses. Similar books have appeared since then – quite apart from whether this thesis is valid or not, there is intrinsic interest in at least reading what they have to say. I should say that 'Stonehenge Decoded' is quite a difficult and complicated book!

[107] approximately 18 years 11 days, depending on how many leap years there have been in the meantime

showing stars (conditions were slightly cloudy such that the large-scale features of the eclipse could be seen but 'stargazing' was obviously diminished) and I think it is probably more accurate to say that any conclusions were uncertain. A parallel team had gone to Sobral in Brazil to observe.

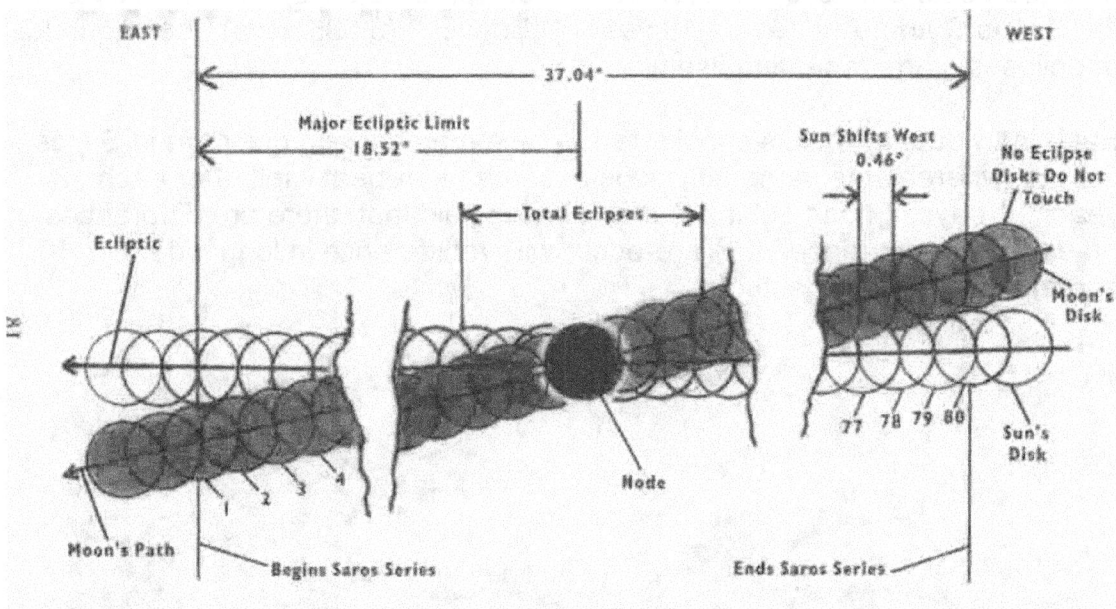

Figure 28 - Schematic of the full Saros Series 136

The above diagram attempts to show how a Saros Series progresses. The saros cycle of 6 585 days relies on the observation that 223 synodical months equals 19 eclipse years[108]. But the correspondence is not totally exact – it is out by what corresponds to an angular distance of 0.46°, so to a higher level of accuracy the eclipses in the series will vary slightly from one occurrence to the next. In Figure 28, the first eclipse will cast a shadow at the southern polar regions. As the Series progresses the eclipse will move northwards. After eclipse number 80, the Series finishes.

This particular series has the maximum possible danger zone (37.04°) and its 80 eclipses will produce a total lifetime for the series of

80 times 6,585 days = 1,442 years

A series with the minimum danger zone will last 1206 years.

Figure 29 shows a 3-dimensional representation of Saros Series 145, which contains both the eclipse of 11th August 1999 which could be seen

---

[108] This is stating the situation simply, but it relies on at least one other coincidence as well. The eclipse year is actually 346 days long and is shorter than 365 days because the 'line of nodes' drawn between the two nodes is itself rotating, in an opposite direction to the sun so that it takes less than 6 months for the Sun to travel from one node to the next.

Figure 29 - Saros Series 145

in Cornwall, and the eclipse of 1927 which could be seen in North Wales and North England.[109]

The fact that they could both be seen in Britain is due to 'overlapping' rather than them being part of the same 'triple saros'. The eclipse of 1927 moved in a 'north-east' direction towards Siberia whereas the 1999 eclipse moved 'south-east' towards central and southern Europe. Both suffered from the clouds but 1927 lasted 20 seconds at most, whereas 1999 lasted just over 2 minutes at most.

Saros 145 is progressing southwards and has been given an odd number, in line with the convention just mentioned.

The length of an eclipse depends on two things - the size of the shadow cast by the Moon and the 'velocity' of the observer. An observer on the equator is moving faster than elsewhere on Earth in the same way that someone on the edge of a roundabout is moving faster than someone closer to the center. Since an equatorial observer is also moving in the same direction as the Moon' shadow, s(he) will tend to view an eclipse for longer than elsewhere on the globe.

The theoretical maximum duration of an eclipse is 7m 31 s and the maximum width of shadow is 270 km. This shadow moves at around 3000 km/h.

The dappling effect observed in woods is due to the 'pinhole camera' effect of the trees' branches and twigs, projecting images of the Sun. During an eclipse these projected images will show the partial phases of the eclipse.

Terminology used to describe the development of an eclipse is as follows

First contact : the Moon starts to encroach on the Sun

---

[109] For completeness, I should state that there was also a totally separate total eclipse 'recently' visible in Britain - in the North of the Shetland Islands in 1954

Second contact : the Moon covers the Sun completely. Immediately prior to this, *Baily's Beads*[110] can be observed as light shines through valleys on the Moon, culminating in one last *Diamond Ring* effect
Third contact : The Diamond Ring and Baily's Beads occur in reverse order as the total eclipse ends
Fourth contact : The Moon moves off the sun completely

Be warned that a 98% or a 99% eclipse is a totally different thing from a total eclipse. The effect will be more like twilight and be generally 'disappointing'.

---

[110] Francis Baily observed the eclipse of 1836 in Scotland, and several others– although he was not the first to observe 'Baily's Beads'. Knowledge was quite primitive at this stage. The first photograph of a total eclipse was taken in 1851, and from this same eclipse scientists were able to pronounce for the first time that the corona and chromosphere were solar phenomena, i.e. nothing to do with the Moon.

Chapter 6   Solar System

Formation of Planets

The nature of the revolution and rotation of bodies in the Solar System implies a common factor in their origin. The most common theory for the formation of the Solar System envisages it forming from a 'solar nebula'. Superficially this is similar to the theory put forward by the great French mathematician Laplace around 1800 but the actual mathematical details differ.

Under the 'current' theory, three stages then follow

1. particles 'stick together' forming **planetesimals**

2. Once planetesimals reach a diameter of 10km then gravity kicks in, initiating a period of 'runaway growth' and 'heavy bombardment'. Collisions result in the formation of larger bodies called **planetary embryos**

3. Collisions continue but at a slower rate resulting eventually in the present-day **planets**

It should be noted that the original solar nebula will contain large amounts of hydrogen and helium and the larger planetary embryos of the outer planets would have been able to 'keep hold' of some of this gas. There is a stage in the early history of stars called the **T-Tauri** stage which is characterized by an enormous outflow of solar wind. This solar wind would have been able to completely blow off any early atmospheres of the 'terrestrial' planets but not those of the outer planets.

Although I might tend to refer primarily to the inner terrestrial planets when talking about various processes in the initial sections of this chapter, in many cases these processes can be generalized to the outer section of the Solar System.

There are, as to be expected, problems with such a theory. For starters, any theory needs to explain why 99.8% of the mass of the Solar System is contained in the Sun, but less than 1% of the angular momentum.

Heating

As you probably already know, the interior of our planet is hot, indicating it has had a fiery past. In fact, the theory says that at one time the entire planet was molten and, in general, this is a phase to be expected in the history of all planetary bodies. This heat stems from three sources in the main

    1. <u>Accretion</u>, i.e. the heat of impact with another body. There is a general rule in physics that all energy eventually ends up as heat, and a significant part of any impact energy does indeed end up as heat.

2. Core Formation During the molten phase, iron sinks to the center. The potential energy lost by material as it drops in a gravitational field is equal to

mass x acceleration due to gravity x height dropped

and although the displaced rocky material will actually require a **gain** of potential energy to rise up from the center, since the mass of the incoming metal will be greater than the mass of the displaced rock there will be a net loss of potential energy – which will tend to increase the temperature.

3. Radioactivity from naturally-occuring isotopes which, although slow in comparison with those used in power stations, will still build up an appreciable store of heat over time.

There is an additional effect which is important for some satellites - **tidal heating**. As has been mentioned already, the Earth's orbit is an ellipse. This will be true for the other planets also, as well as for satellites orbiting planets. These satellites, just like our own Moon, will tend to get locked into a 'stationary' orbit with the same face always turned towards the planet. A tidal bulge will be formed on the satellite and if the orbit was a circle, this bulge would stay fixed. However its elliptical orbit will cause the bulge to 'wander' to and fro slightly, generating friction and thus obviously generating heat. The most famous satellite affected by this effect is Io, the innermost of the large Galilean satellites of Jupiter[111]. It exhibits widespread volcanic activity by virtue of this tidal effect – otherwise its small size (similar to our Moon) and distance from the Sun would leave it a very cold body indeed.

Differentiation

As a result of the molten state of the Earth (and other bodies) differentiation, or layering, will take place. The **crust** that forms initially will be of a different composition (formed of less dense material that floats on top of the molten Earth) in comparison to rocks at a lower level in the **mantle**. The formation of an iron core is also a differentiation effect.

Astronomers generally favor a different classification based on the physical nature of the rocks. In this system the outer layer is called the **lithosphere,** which consists of the crust and an upper layer of the mantle. Below the lithosphere is the **aesthenosphere**. The lithosphere is initially composed of tectonic plates 'floating' on the aesthenosphere[112]. As the body cools, the lithosphere will eventually form into one single rigid layer and plate movement will cease. As cooling progresses further, the lithosphere will then become thicker and thicker.

---

[111] Io, Europa, Ganymede, Callisto – the four most visible Moons of Jupiter whose discovery is usually attributed to Galileo
[112] i.e. in continental drift, it is the lithosphere that forms the plates

An important point worth mentioning here is that any core must be liquid in order to produce a magnetic field. The inherent motion of the core plus the rotation of the planet combine to produce the magnetic field.

How can heat escape?

So how does the internal heat escape from a body, given that heat tends to transfer from regions of high temperature to regions of lower temperature?

The most obvious answer would be volcanoes. Broadly speaking, these come in two different types

- <u>Effusive</u> 'Straightforward' outflowing of lava. This outflow will produce shield volcanoes like those of Hawaii. Basalt, a product of this form of volcanism, covers 70% of Earth's surface and similarly for Venus and Mars

- <u>Pyroclastic</u> This is the more spectacular 'explosive' type, such as the recent St Helen's eruption in Washington state, USA. Historical examples include Krakatoa in 1883 and the Vesuvius eruption that caused the demise of Pompeii and Herculaneum. One or two disasters in the ancient world (both real and assumed) are commonly laid at the door of pyroclastic eruptions

Volcanoes on Earth are commonly sited adjacent to a tectonic plate boundary. Plate movement will cause volcanoes to move away from their source of heat and become extinct (Snowdon is actually an extinct volcano, for example). For a planet like Mars which has no plate tectonics, a volcano will 'stay put' and consequently shield volcanoes can grow to an enormous size – Mons Olympus on Mars is 27 km high (cf. Everest at 8850m, or indeed Hawaii at about 8000m from their bases on the sea bed, which is particularly relevant in comparing with Mars because the height of Martian mountains is not measured 'above sea level').

Other factors do need to be considered. For example, the lower gravity of Mars reduces the weight of volcanoes - easing the load that the surface needs to bear, whereas Hawaii has noticeably 'sunk' under the weight of its volcanoes. In this sense the surface of Earth is not strong enough to tolerate very high mountains.

Ice volcanoes are possible on satellites of the giant planets.

Atmospheres

As I have just mentioned the early atmospheres of the terrestrial planets would have been lost during a T Tauri phase, which would also have ejected any interplanetary hydrogen and helium elsewhere in the Solar System.

Any current atmospheres of the inner planets will have originated from outgassing – gas 'seeping out' from the planet's interior. Differences in atmospheres are caused by different local conditions, so that although Carbon Dioxide predominates on Mars and Venus, this gas is reduced on Earth by various factors such as interaction with water (the oceans), or by carbon becoming 'tied up' in living bodies and in resultant rocks such as limestone. It is also absorbed by living plants which 'exhale' oxygen, the source of that particular gas on Earth.

Carbon Dioxide and Water Vapor are greenhouse gases, i.e. they tend to absorb infra-red radiation that would otherwise exit to space. This does have some beneficial effects – without it the Earth would have an average temperature of 255K (-18°), instead of 288K (+15°).

Atmospheres are typically layered. The lowest layer on Earth is the troposphere (where all the weather is - aircraft tend to fly above this, if they can). It is heated by infra-red radiation given off by the Earth's surface, not by the direct absorption of sunlight

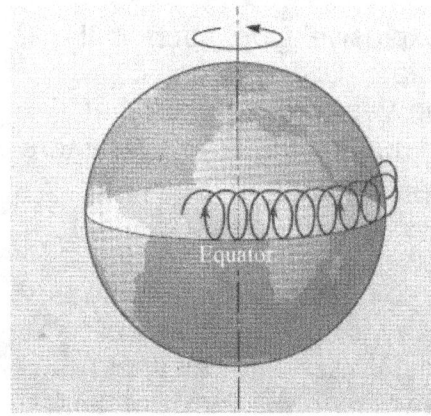

Figure 30 - Hadley Cell, north of equator

Hadley Cells are convection cells. Figure 30 shows such a cell formed from air rising at the equator. As it moves North, it will be carried eastwards of the longitude it had at the equator, by virtue of the fact that the ground speed reduces as you move away from the equator but the air will still be moving with roughly the same eastward velocity that it had at the equator. At about 30° north of the equator it will descend and tend to return to its starting point, except that frictional losses etc. will mean that it will not return to its original starting point and will thus, as shown in the diagram, spiral within fixed latitudes.

It is these Hadley Cells that produce the trade winds. In the Northern Hemisphere, trade winds blow from North East to South West; and in the Southern Hemisphere from South-East to North-West[113].

The spread of a Hadley Cell latitude-wise is governed by the planet's rotation. As shown in Figure 31, the Earth has three such cells in each hemisphere. Venus, which rotates very slowly will only have one cell in each hemisphere, spreading from the Equator to the poles. Jupiter, which rotates once every 10 hours will have many cells – which are

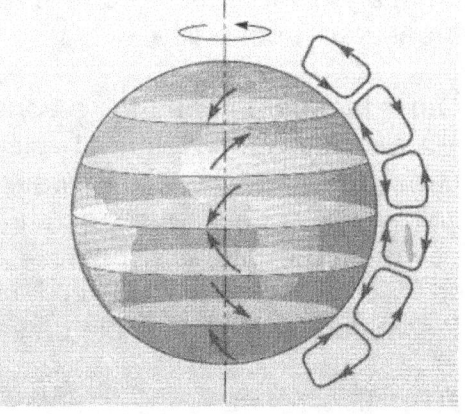

Figure 31 - 3 Hadley Cells in each hemisphere

---

[113] the winds can be affected by mountains and land formations

responsible for producing the white zones (air rising producing clouds) and the dark belts (air descending).

Noctilucent clouds are objects of interest to some astronomers. These are five times higher than normal clouds and are thus high enough to reflect sunlight late into the night. They only form during the summer and their cause is unknown – they could be due to water vapor ejected by volcanoes or even be due to pollution. They are invisible in day time.

Craters

Cratering is common throughout the Solar System and the Earth has not been exempt from meteorite hits. In recent times the Chicxulub crater has been uncovered in the Yucatan Peninsula in Mexico under about 1 km of sediment, and the meteorite strike which produced it is becoming more accepted as the cause of the dinosaur extinction about 65 million years ago. Its size is a bit uncertain because it is buried under about so much sediment, but current estimates are between 120-180km.

Most known craters on Earth are of 100m diameter and upwards, but obviously many have been weathered away, or degraded such that are not instantly recognizable from the ground. In some instances they have become lakes.

The Meteor Crater in Arizona is a well-known impact crater on Earth, although it was caused by a meteorite, not a meteor. It is also known as the Barringer Crater.

Figure 32 - Meteor Crater, Arizona

The impactor is now known to have been of iron which mostly vaporized in the impact. It was of the order of 50m in diameter. The crater itself is about 50 000 years old, and is of 1.2 km diameter. In comparison with lunar craters, this is quite tiny.

When you think about it, it is not immediately obvious why craters should be circular – this is in fact due to the explosive impact, explosive in the real sense of the word. Craters do show some variation, e.g. Martian craters show the effects of water on the surface at the time of impact. Ironically, craters on icy bodies look more like lunar craters than Martian ones – at extremely low temperatures ice resembles rock.

Ice does tend to relax more easily, although in such cases the imprint of the crater could still remain, as what is called a palimpsest.[114]

Classification of Solar System Bodies

In 2006, the International Astronomical Union issued a new classification for objects in the Solar System – there are three classes

1. Planets
2. Dwarf Planets
3. Small Solar System Bodies

Loosely speaking, a dwarf planet is classified as a near-spherical body[115] which, contrary to a 'real' planet, has not 'cleared its neighborhood', i.e. it has not become the dominant body within its orbital region. At the time of writing, there are five dwarf planets – in order of size : Pluto, Eris, Makemake, Haumea, Ceres,

The status of Pluto is further complicated by the fact there is no standard definition, as yet, of what constitutes a double planet system.

In 1995, the first two objects of what was to become the Kuiper Belt were discovered. This is an area further out than Neptune, from 30 to 50 AU, where there are estimated to be about 100,000 objects of larger than 50km radius. The Kuiper Belt is also assumed to be the source of short-period comets, which tend to orbit close to the plane of the Solar System (the Oort Cloud is now considered to be the origin of long-period comets, which typically orbit outside the plane of the Solar System).

Brief review of Planets

Mercury

Mariner 10 is the only probe that has visited the planet (in 1974). All other things being equal, you might expect Mercury to keep the same face to the Sun. Only in 1965 was it observed that its rotation rate is 56.65 days[116], which is exactly 2/3 of its orbital period of 87.696 days – a state of affairs which is attributed to the high eccentricity of its orbit[117]. This means the solar day on Mercury is 176 days long.

The temperature varies between +430° to -180°. The planet has an iron core and a magnetic field

---

[114] Originally palimpsests were manuscripts where the medium has been re-used and you can still see evidence of the original writing underneath, despite the attempts to erase the original writings.
[115] The object would have to reach a certain size before gravity was able to produce a near-spherical shape. SSSBs tend to have noticeably non-spherical shapes
[116] This was observed by radio telescopes using radar
[117] its eccentricity is 0.2056

As seen in the sky, its elongation (its angular distance from the Sun) has a maximum of 28°.

Venus

It has a sidereal rotation period of 243 days and an orbital period around the Sun of 225 days. Furthermore, it rotates in a retrograde direction (i.e. clockwise when viewed from the North, from where the Earth and most other objects appears to rotate counter-clockwise) – so that the solar day is 116 days long[118].

Although its mass and diameter are similar to Earth, general conditions are markedly different. Its atmosphere is mostly CO_2 (96%). The surface temperature is maintained at about 480° day and night. Despite wind speeds at the surface being low, speeds higher up become significant. The planet is covered with thick clouds formed from 75% Sulfuric Acid and 25% water. It has no magnetic field, presumably because the planet is rotating so slowly.

The planet has been subjected to a runaway greenhouse effect – an initial greenhouse effect has caused evaporation of water, and this additional water, being also a greenhouse gas, has caused even more evaporation, and so on. The water in the atmosphere has then been dissociated by Ultra-Violet into hydrogen and oxygen, and the hydrogen has tended to escape from the atmosphere completely.

It has been necessary to map the planet using radar, the most recent being by the Magellan probe in 1990-1992. There are vast plains with some depressions and mountains, most of which have limited depth/height. Two prominent upland areas are called Ishtar Terra and Aphrodite Terra, both with a size comparable to a continent on Earth.

Both Mercury and Venus can transit the Sun. Transits of Mercury are quite rare but transits of Venus are even rarer. They occur about every hundred years and then come in pairs, e.g. 1874 and 1882, 2004 and 2012. In earlier times, transits attracted some interest because it was theoretically possible to use them to calculate the Astronomical Unit – one of James Cook's voyages to the Pacific was carried out for the purpose of observing such a transit.

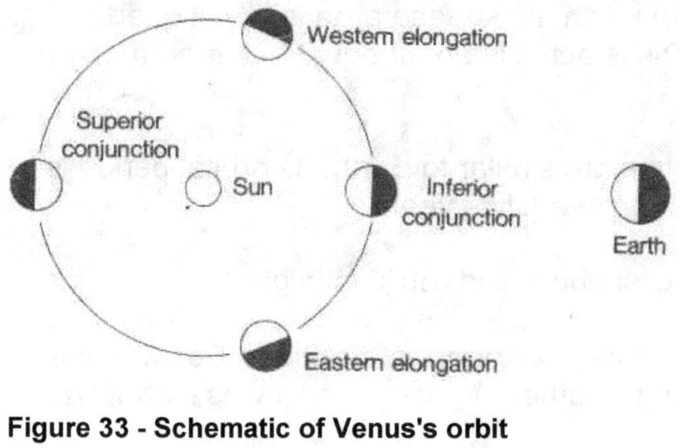

Figure 33 - Schematic of Venus's orbit

Venus is obviously never at opposition, but it can be at either inferior conjunction or superior conjunction.

Its elongation (distance from the Sun) is never more than three hours of Right Ascension. It appears as

118 on most planets, the solar day is longer than the sidereal period, but because Venus is rotating in a retrograde fashion, the solar day is shorter than the sidereal day.

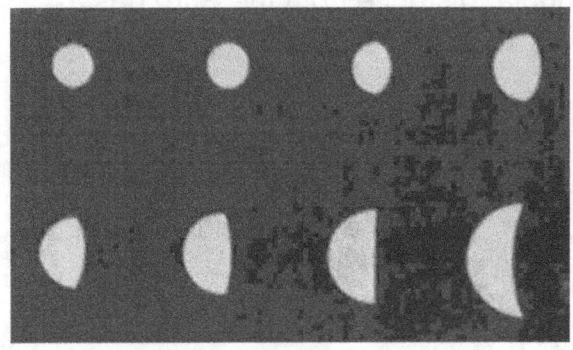

Figure 34 - Phases and relative sizes

either the Evening Star or the Morning Star. The cloud covering produces an albedo of 0.76, the highest albedo in the Solar System

The phases were first recorded by Galileo (and were strong evidence in favor of the Sun-centered planetary system). The planet is not brightest when full because then it is at superior conjunction and at its farthest point away from us. Venus is brightest when it is slightly less than half-illuminated (the largest-disk in Figure 34 - **Phases and relative sizes**). Then it is at a magnitude of about -4.4 (15 times brighter than Sirius). Also you can see from Figure 33 that it is <u>not</u> at half phase (dichotomy) when it at '90°' to us.

It can be seen during the day. In my experience if you view it as The Morning Star then you can still see it as dawn breaks and after, but if you should lose track of it that it is very hard to locate it again.

Earth (briefly)

The Earth has an iron core – to be more precise a solid inner core and a liquid outer core (a liquid core being required for the production of a magnetic field, which acts as a protective shield for the Earth)

The oxygen in its atmosphere is not a product of outgassing, but of biological action by plants.

Craters smaller than 15 km have eroded away and craters above this have degraded. In general, they have no ejecta blankets, raised rims or central peaks.

Mars

Its mass is about 1/10 the size of Earth. Its surface area is about quarter that of the Earth but this surface areas is actually about equal to the area of the land mass on Earth.

Its rotation period[119] and inclination are similar to Earth. Its orbital period is 687 days and is at opposition about every two years.

The temperature can reach 10°C at noon, and -60°C at night

It has a very elliptical orbit in astronomical terms (at Southern Summer it is 20% closer to the Sun than during Southern Winter), and it was by analyzing

---

[119] Its rotation period is 24 hr 37 min, so observers tend to see almost the same face night after night.

the motion of Mars that allowed Kepler to produce his first law, that planetary orbits are ellipses – not circles.

Its thin atmosphere is mostly CO_2[120], although it must have been thicker in the past, allowing liquid water to exist of which there is strong evidence in the form of dried-up river beds. The atmosphere would have been lost because of the low gravity and the cessation of outgassing, although many believe the actual water is still there, e.g. under the surface as ice.

Thanks to space probes, we can see that Mars has craters, volcanoes (one of which, Mons Olympus, is the highest volcano in the Solar System) and deep canyons (one of which, the Mariner Valley, is the largest canyon in the Solar System).

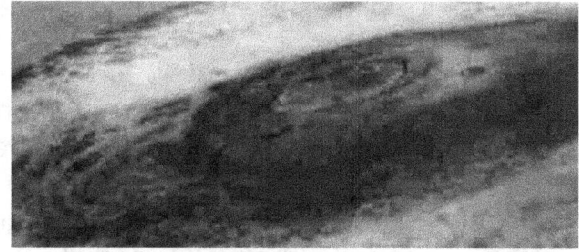

Impact basins include the Hellas Basin (dia. 2000 km) and Agyre Basin (dia. 1200 km). There is an absence of small craters due to weathering and the larger ones could be degraded by the same process.

Figure 35 – Olympus Mons

Overall, about 2/3 of Martian probes have failed before the planned termination date. As of the end of 2018, about 50 launches have been attempted. In 1999, the Mars Climate Observer was lost because of a classic mix-up between metric and imperial units.

The first lander was Viking 1 of 1976, coupled with Viking 2 soon after.

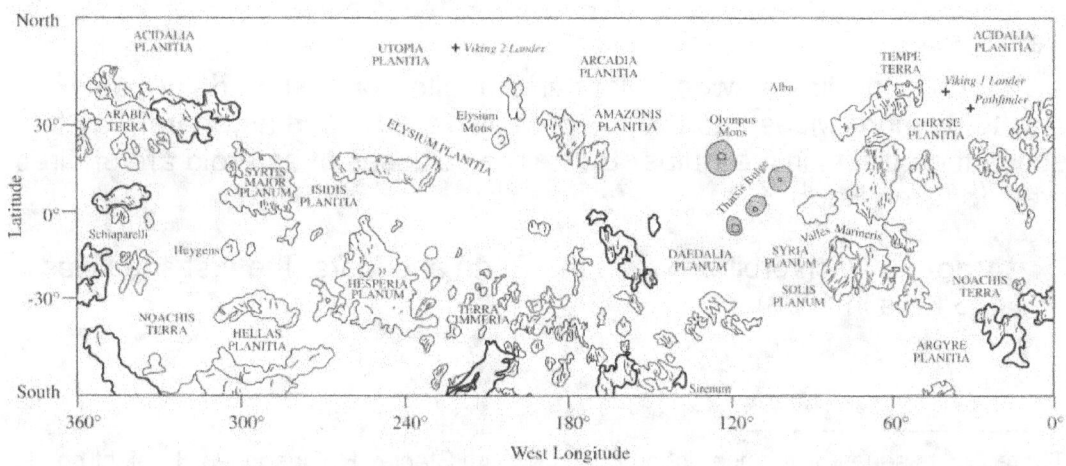

Figure 36 - Mercator projection of Mars

The two hemispheres of Mars are noticeably different – the North tends to be low-lying plains while the South tends to be dominated by highlands, which

---

[120] pressure on Mars is about 6 mm (Earth's being about 1000 mm)

are more heavily-cratered and therefore older[121]. There are exceptions – the Hellas impact crater in the South contains the lowest point on the planet while the North has several lofty volcanoes[122]. The Northern plains have been re-surfaced by volcanic emissions, largely erasing any craters.

On the map above, plains are designated as 'planitia' and plateaus as 'planum'. The Mariner Valley[123] and the three volcanoes of the Tharsis region are shown in Figure 37.

In the late 19th century/early 20th century, study of the planet seems to have been bedeviled by a belief in Martian canals. E.E. Barnard seems to have been reticent about publishing his discovery of craters for fear or ridicule. On the other hand, there is a story to the effect that when asked whether he could see canals, he replied that unfortunately his telescope was too powerful to see them.

Figure 37 - Mariner Valley, with Tharsis at left

For observers on earth, ice caps are noticeable at the poles (composed of both CO_2 and H_2O) which ebb and flow with the seasons. Changing clouds/dust storms can also be observed which sometimes encompass the entire planet. Dust clouds can block the view even in the most powerful telescopes, and even in the most powerful telescopes no feature of less than 300 km across can be viewed under clear conditions.

It has two small satellites, Phobos and Deimos, which are usually considered to be captured asteroids.

Recent probes include

Asteroids

The 'Asteroid Belt' lies between Mars and Jupiter, but asteroids are now known to be more widespread. You could also the term in a general way to describe the objects in the Kuiper Belt. Three classes of asteroid are of direct interest to us on Earth

- **Amors**, which orbit between the Earth and Mars (the first seen was 433 Eros in 1898)

---

[121] There are as series of science-fiction books by Kim Stanley Robinson which might be of interest in the context of Mars. Red Mars, Blue Mars and White Mars relate events on a terra-formed Mars of the future, in which the Northern Hemisphere has mostly become an ocean. These stories have brought Robinson the prestigious Hugo Award (an award for science fiction).

[122] Volcanoes are concentrated in the Tharsis and Elysium regions.

[123] A canyon, much larger than the Grand Canyon (capable of stretching from one coast of America to the other), formed by faulting rather than by the action of water. Nevertheless it does show signs of weathering and water could also have flowed along it. It is named after the Mariner 9 space probe. More exactly, it is a network of canyons (a canyon on Mars is often referred to as a chasma)

- **Apollos**, which also orbit between Mars and Earth on the whole, but can cross the Earth's orbit during their elliptical orbits (the first seen was 1862 Apollo in 1932)

- **Atens**, which orbit between Earth and Venus on the whole, but which can also cross Earth's orbit at aphelion (the first seen was 2062 Aten in 1976)

Although some of these asteroids can nominally cross the earth's orbit, they tend to orbit at large angles to the plane of the Solar System (up to 60°), so a 'crossing' can occur when the two bodies are actually a significant distance apart.

Asteroids are numbered in order of discovery, as used above.

The Galileo probe to Jupiter made close contact with a couple of asteroids on its journey to Jupiter, namely.

- 951 Gaspra
- 243 Ida, which was found to have a satellite, Dactyl. Ida is a highly irregular shape but its longest length is 52 km long.

The NEAR (Near Earth Asteroid Rendevous) spacecraft fly past Mathilda in 1997 and went into orbit around Eros in 2000, landing the year after.

In 2005, the Japanese probe Hayabusa landed on 25143 Itokawa and returned samples to Earth. This was followed by Hayabusa2, launched in 2014 to 162173 Ryugu which it reached in 2018. It is intended to again collect samples for return to Earth in 2020.

Asteroids are grouped into several classes, which are closely connected with meteorites as regards composition, e.g.

C class – has a low albedo with the same composition as a class of meteorites known as carbonaceous chondrites
S class – has high albedo and are stony (made from silicates)
M class – are rich in metals

The influence of Jupiter is considered to be such that on the whole more collisions in the asteroid belt were destructive than constructive, thus hindering the growth of a planet in this particular region.

Jupiter

Jupiter is 1300 times the size of the Earth and has a mass 318 times that of Earth's. It is two and a half times as massive as all the other planets put together. The 'surface' has horizontal markings - belts (which are dark) and zones (which are light). The zones are actually clouds produced at the top of a Hadley Cell and the belts denote where the gas is descending again. Any additional coloring is due to trace elements.

Its interior is believed to have three layers : from the center outwards - rocky core, metallic hydrogen layer, liquid molecular hydrogen layer (which is overlaid by the 'atmosphere', which is 90% hydrogen).

Its rotation is about 10 hrs (shorter than other planets, although only just). This produces a departure from a spherical shape, i.e. polar flattening and bulging at the equator, which even a small telescope will detect.

The Great Red Spot is a famous feature which was probably seen by Robert Hooke in 1664 and Cassini around the same time, although it has only been seen with certainty since 1830, and has only been 'red' with certainty since 1878. Its longest diameter is greater than the diameter of any terrestrial planet (25,000 km by 12,000 km, an elliptical shape). Its long existence is due to the fact that there is no solid surface to dissipate the energy although its size has diminished. It is not unique as such – it is the largest of 'hundreds' of storms on the surface

The magnetic field is produced by the metallic hydrogen[124], although the 'South Pole' is aligned with geographic North, i.e. a compass would point towards geographic South (The Earth itself has reversed its polarity at least nine times during its lifetime).

The first space probe to investigate Jupiter was Pioneer 10 (arrived December 1974). The latest probe was Galileo which went into orbit around the planet in December 1995. The mission also investigated its satellites and the mission was ended by sending the probe spiraling into the planet itself[125].

---

[124] It is enough to accept at this stage that this is hydrogen under enormous pressure which acts like a metal
[125] Earlier, it had released a smaller probe into the planet, which managed to survive for an hour, reaching about 150 kilometers below the visible cloud tops..

For observers it is well placed for several months a year. Its orbital period is 12 years, so it will move one constellation a year[126]. At opposition it will typically have a magnitude of -2 to -2.5.

Satellites of Jupiter

Io is the nearest of the Galilean satellites to Jupiter and is a rocky world. Europa has an icy surface but is mostly rock. Ganymede and Callisto have progressively thicker icy coverings. Callisto might not have fully differentiated – its core could be a mixture of rock and ice. In several respects, the system is like a Solar System in miniature

- **Io** quite apart from its distance from the Sun, the small size of Io would otherwise lead to it being a very cold location – if it were not for tidal heating producing much volcanism. Even then, temperatures of -176°C can be recorded in some locations.

- **Europa** covered in nearly pure water ice, the smoothest body in the Solar System. The surface tends to suggest 'ice tectonics', with the ice breaking into blocks now and again and 'shifting around' and with the surface regularly being renewed. There is assumed to be a lake of water below the ice

- **Ganymede** This is the largest satellite in the Solar System. Galileo (the space probe) detected a magnetic field, which no other satellite has. This tends to be explained in terms of an iron core which is still circulating for unusual reason(s).

- **Callisto** the most cratered object in the Solar System, the surface hasn't been altered for years.

	Mean Radius (km)	Orbital Period (Earth days)	
Io	1820	1.77	volcanic
Europa	1570	3.55	smoothest body in SS
Ganymede	2630	7.15	largest moon in SS
Callisto	2400	16.69	most cratered body in SS

You can see above that there is resonance between the orbits – Europa's period is twice that of Io and Ganymede's is twice that of Europa. This will result in regular 'jolts' to the orbits, keeping them elliptical. Otherwise they would become circular under the influence of Jupiter's gravity.

- Of the smaller satellites discovered, several orbit in a retrograde fashion and are presumably captured asteroids. Of these satellites,

---

[126] Jupiter travels through 12 Chinese constellations, one per year - and a particular year is named after the constellation in which Jupiter lies – during the Year of the Dragon, Jupiter lies in the constellation of the dragon, etc.

those whose name ends in an -a orbit in a prograde fashion (i.e. in the same direction as the rotation of Jupiter) and those that end in –e orbit in a retrograde fashion.

Saturn

Saturn has prominent rings possibly due to a break-up of 'large' satellite. Huygens was first able to recognize the nature of the rings, using a telescope he had made himself. It is now known that all the outer planets have rings which are obviously much less prominent

The rings are denoted alphabetically in their order of discovery, so the sequence going outwards is

D,C,B,A,F,G,E

The Cassini Division is a well known feature separating the A and B rings. The orientation of the rings varies. Sometimes, when they are edge on, they become hard to see – this happens about every 15 years (about half of the planets orbital period which is around 29 years).

Its average density is about half that of Jupiter. Like Jupiter, it is flattened at the poles by high rotation. The composition of its internal layers are similar to Jupiter also. Further the surface is also banded.

A White Spot was famously sighted by Will Hay[127], and similar spots have been seen since, possibly due to ammonia clouds in the atmosphere.

The most recent space probe to visit is Cassini-Huygens, which arrived in 2004.

Titan
Titan is the sole satellite in the Solar System to have an atmosphere – it is nitrogen-rich and the ground pressure is 150% that of the Earth's atmosphere.

Uranus

Uranus was only discovered in 1781, by William Herschel. Very little was known about it until visited by Voyager 2 in 1986.

It is strongly tilted – by 98° so that technically it is rotating in a retrograde fashion. Its magnetic field is strange in being angled at 55° to the axis of rotation although this places the magnetic field at a more 'usual' angle with respect to the plane of the Solar System, so the magnetosphere is not unusual[128].

---

[127] Will Hay, comedian and astronomer
[128] i.e. the magnetic field is not unusual in the way that it interacts with the charged particles of the solar wind.

Unlike other giant planets, it is not emitting heat. Other giants emit more energy than they receive from the Sun. Doubts have been expressed in some circles as to whether the planet has any sort of rocky core.

Despite being on its side, all its satellites orbit in an equatorial plane. The satellites break with the tradition of using names from Greek and Roman mythology and are named mostly after Shakespearian characters (with a few named after characters from Alexander Pope). Titania and Oberon are its largest satellites. Its 'fifth' satellite, Miranda, was only discovered in 1947 and has often been described as looking as though someone has pulled it apart and put it back together again.

Neptune

Neptune has been seen by Voyager 2 to have a Great Dark Spot (similar to the Great Red Spot at first sight, but dissimilar in details), although it disappeared soon after .

It receives 3% of the sunlight that Jupiter does and has 5% of Jupiter's mass, but has a mass similar to that of Uranus[129]. Uranus and Neptune can be considered in several ways to be 'twins'. Like Uranus it has no metallic hydrogen, so its magnetic field is assumed to come from some liquid layer containing ions. This is probably why the magnetic field is displaced from the center of the planet – it is also inclined by 47° from the axis of rotation, consistent with the behavior of Uranus.

Both Uranus and Neptune appear blue because methane in their atmospheres absorbs red light. Both planets have a much higher proportion of icy materials - water ice, ammonia ice and methane ice. Differentiation is possibly incomplete

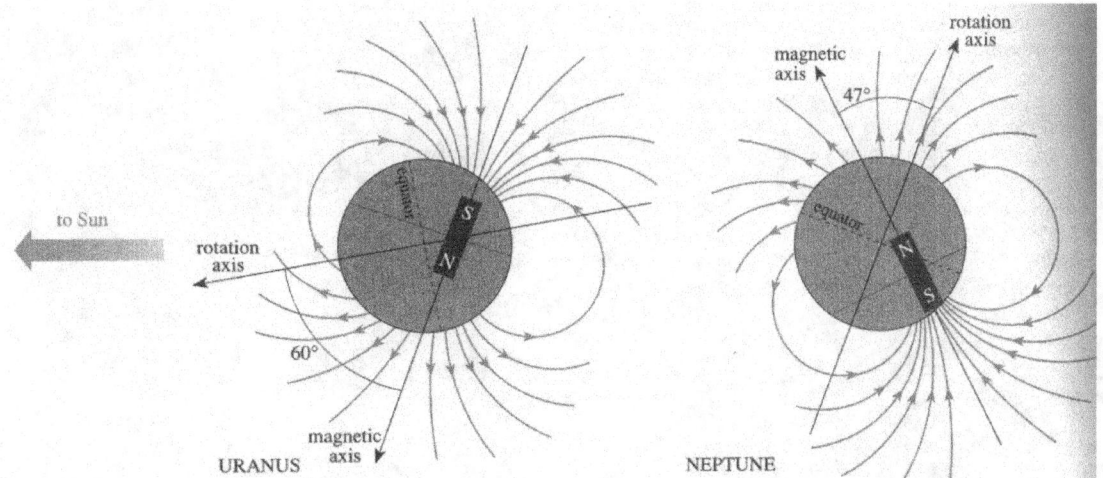

The largest moon is Triton and this is the only large moon to revolve in a sense opposite to its planet's rotation (i.e. in a retrograde manner)

---

[129] Uranus : 17 Earth masses. Neptune: 15 Earth masses

Chapter 7   The Moon

The Moon has a diameter of 3500 km and its surface is divided between the so-called 'maria' or 'seas' (actually lava-filled basins) and the 'highlands'. The maria are a couple of kilometers lower than the highlands. It is quite obvious that the highlands are more densely cratered than the maria, but the maria are most certainly not young – they are maybe 3.2 - 4 billion years old[130] compared with the age of the entire Moon of maybe 4.5 billion years. The situation is at least consistent with the theory of solar system formation presented right at the beginning of the last chapter, where 'stage 2' consisted of a period of heavy bombardment as planetesimals became large enough to interact gravitationally and eventually produce planetary embryos. The maria seem to have been formed after these events.

Mapping

In March 1610, Galileo produced his *Sidereus Nuncius*[131] detailing, among others, his Moon observations of 1609. This book had enormous influence and also attracted enormous opposition.

Since I am based in Britain, I cannot help but be aware of rival claims for the first telescope observations of the Moon. Thomas Harriot observed the Moon on 26 July 1609. On later viewing Galileo's book, he attempted his own drawings although these seem to have had little influence (this is an assumption on my part because these drawings were only rediscovered in 1965).

Following on from Galileo's discovery, there appears to have been some competition between a few astronomers and / or cartographers for the honor of producing the first map (Langrenus and Hevelius are particularly important names in this respect), something that went hand-in-hand with improvements in telescopes. As it happens, the main features on the Moon derive their names for the most part from a map in the 'New Almagest' by Giovanni Riccioli which appeared in 1651 (seemingly based largely on observations by his pupil, Grimaldi). The work as a whole was originally perceived as an anti-Copernican tract[132] (it mentions Galileo's ideas as having 'been condemned as heresy, or at least as erroneous') but nevertheless the decision by Giovanni Cassini, the director of Paris Observatory, to use the nomenclature was highly influential in its universal adoption.

---

[130] the seas differ in age

[131] this can be translated as either 'Starry Message' or 'Starry Messenger'. Galileo might have meant the first but the second interpretation seems to have landed him in hot water with third parties later on

[132] and contained a record of Galileo's official sentencing by the Inquisition in 1633. This became important because the original was among the Vatican records carried off to Paris by Napoleonic troops in 1814 and was found to be missing when the records were later returned.

Hieronymus Schröter was the first person to map details like rilles. He added about 70 new names. Tragically, all his equipment and records (in Lilienthal, nr. Bremen) were destroyed by French armies in 1813, which brought his work to an permanent end .

Johann Heinrich Mädler[133] produced an authoritative map which appeared in 1837-8 and was about a meter square. According to the 'Guinness Book of Astronomy' this was the 'first really good map of the Moon'. He was the first to state authoritatively that the Moon was a lifeless and unchanging world. He also produced a book 'Der Mond' which was well-respected but has never been translated into English.

Developments since then have relied on photography, which can 'capture' features more easily than the eye. Latterly the moon has been mapped in more detail by the Clementine and Lunar Prospector probes.

Due to a decision of the IAU (allegedly under some pressure from the space agencies), the 'left' of the moon is now designated West and the 'right' is East. Previously this was the other way around and has the effect that the 'Oriental Sea' is now in the West.

Metonic Cycle

This is a cycle of 19 years after which the phases of the moon start repeating themselves on the same dates of the solar year[134]. Like the Saros Cycle, this was known to the ancients and was well-known by the time of the Council of Nicaea in 325AD which set the date of Easter. Easter Day is the first Sunday after the first full Moon following the vernal equinox.

Maria (Seas)

The seas were formed by lava filling large impact craters. The impact that formed the Mare Imbrium, which has a diameter of 1250 km, came very close to splitting the moon apart. The Mare Imbrium is surrounded by mountain ranges bearing terrestrial names, in clockwise order – Jura, Alps, Caucasus, Apennines, Carpathians. These 'mountains' are really the old crater walls. Individual 'mountains' lie inside the crater which again are probably just a part of the former inner ring structure, e.g. Mt Piton at 2200m. in the north-east and Mount Pico and the Teneriffe Mountains both south of the Plato crater.

---

[133] To be more precise Mädler often worked with Beer, although it is often (perhaps unfairly) assumed that Mädler was the scientist while Beer tended to provide the finance
[134] also known as a tropical year - it is the time that the Sun takes to return to the same position in the cycle of seasons, as seen from Earth; for example, the time from vernal equinox to vernal equinox, or from summer solstice to summer solstice. Because of the precession of the equinoxes, the solar year is about 20 minutes shorter than the time it takes Earth to complete one full orbit around the Sun as measured with respect to the fixed stars (the sidereal year).

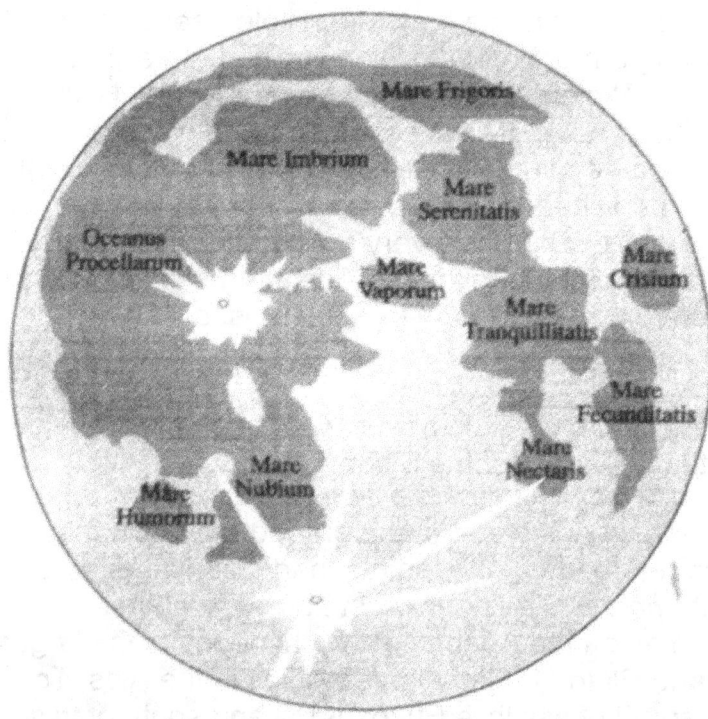

Figure 38 - Maria, plus Copernicus and Tycho

Rocks from Tranquility have been dated to about 3.8 billion years ago, although bear in mind that the lava did not cover the seas immediately but rather oozed out intermittently over about a 100 million years or so.

Mare Crisium looks elongated on the limb but it is actually circular.

All the Moon is covered by regolith – this is a layer of dust which, depending on the location, is from a few millimeters to several meters thick. Over time micrometeorites, solar radiation and the effects of temperature variations have turned rocks to powder. On their return to Earth, the astronauts required a couple of weeks before all the dust washed off them. They described the smell as being similar to gunpowder.

Craters

Including craters of the size of Arizona's Meteor Crater would result in a Moon map with an enormous number of craters, so most craters shown on a typical Moon map are significantly larger.

I have already previously mentioned that when you look at a map of the Moon, there is actually a problem when you think about – why are most of the craters circular? It is actually the explosion accompanying impact that produces the circular shape

The overall shape of craters can be classified according to their diameter, as follows

Approx. Diameter (km)	
Less than 15	bowl shaped
15-140	flat floor, terraced walls, maybe single central peak
140 - 175	clusters of central peaks
175-350	no central peaks, more than one ring
350-	complex multi-ringed structures – often called 'multi-ringed basins'

Almost at the center of the moon we can see the crater Ptolemaeus (diameter 148km, depth 2.4km) occupying a prominent position, which I am assuming stemmed from Riccioli's belief in Ptolemy's Earth-centered universe. His 'opponents' are also there, which I have always assumed to be a compromise – a compromise he felt necessary to make if his system was to be definitively adopted[135], although they are 'banished. to the Ocean of Storms (Oceanus Procellarum). Copernicus (diameter 93km) is a very noticeable rayed crater, but Aristarchus[136] and Kepler are smaller (although Aristarchus is actually the brightest location on the Moon and Kepler (diameter 35km) is also a rayed crater). Galileo is a very minor crater.

An observer placed at the very center of Ptolomaeus would be unable to see the boundary walls, such is the curvature of the Moon.

Riccioli split the moon into various sectors – e.g. the area to the west of the Mare Nectaris has several craters named after saints, particularly the trio of Theophilus, Cyrillus and Catherina[137]. The north-west has several prominent 'Greek' craters. Eratosthenes lies on the Apennines, Archimedes is the largest crater on the Mare Imbrium and Plato (110km diameter) lies in the Alps. To the east of Plato is Aristotle, south of which is Eudoxus[138], and south of this again is Alexander.

The north-east has several names from Greek mythology. On the edge of the Sea of Serenity is Posidonius, while further North we have Hercules and Atlas close to each other.

Going back to the center again, Ptolomaeus lies atop of a trio with Alphonsus[139] and Arzachel[140]. Just north of Ptolomaeus is the small crater of Herschel. To the east lies Albategnius[141], north of which lies Hipparchus. Inside Hipparchus is Horrocks, and on its southern edge is Halley.

Archimedes (95km) on the Mare Imbrium is flanked by the two craters of Autolycus and Aristillus. Within the same field is Cassini, easily recognized by two craters within Cassini itself.

On the Sea of Fertility you can see the twin craters Messier A and Messier B, assumed to be the result of a hit from a double object. Between Fertility and Tranquility lies Taruntius (diameter 56km) which was used by the Apollo 11 astronauts as a marker during landing.

---

[135] but this is just an assumption on my part

[136] a Greek astronomer from 200BC who is often taken to be the first person to propose the Sun-centered planetary system, but generally considered to be ahead of his time

[137] Catherina P is a crater within Catherina – a fairly common notation with the name of the main crater appended by a letter of the alphabet

[138] ancient Greek astronomer. Usually credited with the idea of celestial bodies being attached to transparent spheres

[139] King of Castile who oversaw the drawing-up of the Alphonsine Tables, improved tables of planetary motions (1270)

[140] Arab astronomer from Spain

[141] Arab astronomer from Mesopotamia (9th-10th centuries)

Riccioli and Grimaldi immortalized themselves with two craters on the western limb (Grimaldi is the darkest location on the Moon). North of this pair is Hevelius (sometimes referred to as just Hevel). On the edge of the Sea of Fertility is Langrenus – Langrenus himself seems to have named this very crater after himself on his own map (which failed to be accepted).

The largest crater on the near side is Bailly (295 km, 3.96 km depth) on the southern limb. Easier to view is the second largest crater - Clavius[142] (232km, depth 4.01km) just a bit further to the north-east.

South-Pole Aitken basin is 2700 km wide and 11km deep, the biggest hole in the Solar System. There is a so-called sea on the far side – the Moscow Sea, but this is really just a crater.

Other features

The Sinus Iridium (Bay of Rainbows) is really a flooded crater - the northern edge is about 600 meters lower than the surface of the Mare Imbrium. Le Monnier is a similar feature off the Sea of Serenity.

Rilles are collapsed volcanic tubes, smaller versions of which exist on Hawaii. Schröter's Valley extends 160 km north from Herodotus (to the west of Aristarchus) - and has a feature called the 'Cobra's Head' at this end. Hadley Rille on the Mare Imbrium was investigated by Apollo 15 – it is 132km long, 1-2 km wide and 370m deep. On the Mare Vaporum you can see the Hyginus and Ariadaeus Rilles (Rima Hyginus and Rima Ariadaeus, in Latin).

Shallow-sided domes are a remnant of volcanic activity. There is a concentration near Marius (between Kepler and Galileo).

Nowadays the crust is 60km thick and the solid mantle a further 800km deep. There may be a molten core below this but volcanism is in general not to be expected. Nevertheless, there are reported incidents known as TLP (Transient Lunar Phenomena) which some try and connect to some residual volcanic activity.

Mascons

The name derives from mass concentrations, i.e. areas of enhanced density. They occur only on the near side and are closely coincident with the impact basins

Observing

The maria might reflect less light than the highlands but they also appear darker because of a biological effect similar to that which operates with sunspots. The Moon only has a total albedo of 7% (only 3% for the maria) - it is actually quite dark, one of the least reflective worlds in the entire solar

---

[142] location of the mysterious monolith in '2001, A Space Odyssey'

system.[143].The brightness of half-moon is 1/9th that of a full moon – primarily because of its shadows.

The terminator is the name given to the line separating the illuminated and dark areas. It moves about ½° an hour.

In general it is best to view craters when they produce shadows, i.e. not during a Full Moon. The exceptions are rayed craters of which Tycho and Copernicus are the best examples. These rays are ejecta from the original impact – over time they are eroded by micrometeorites etc., implying that rayed craters are young. Copernicus is estimated to be about 1 billion years old whereas Tycho is much younger at about 300 million years. Tycho's rays are more extensive implying that maybe the Copernicus impact was less violent or maybe the maria material in which Copernicus lies was more flexible than that in the highland area in which Tycho is to be found. Rays from Tycho can be seen on the seas of Serenity and Tranquility. On Serenity, this ejecta passes adjacent to the crater Bessel and astronauts from Apollo 17 on the south-east edge of the sea came across rocks which had been thrown out by the Tycho impact.

Aristarchus and Kepler are also rayed craters, the former being about 450 million years old. Aristarchus is also a favorite location for sightings of TLP (Transient Lunar Phenomena).

The Moon will move its own diameter in the sky in two minutes. On average, it rises about 50 mins later every day – this difference will vary because of the inclination of the Moon's orbit. In early Autumn the difference is at a minimum – you have a full moon called the Harvest Moon rising just as the Sun is setting and then a few following days when the period of darkness between sunset and moonrise is small, allegedly helping the farmers gathering their harvest. The following full moon is called the Hunter's Moon for analogous reasons.

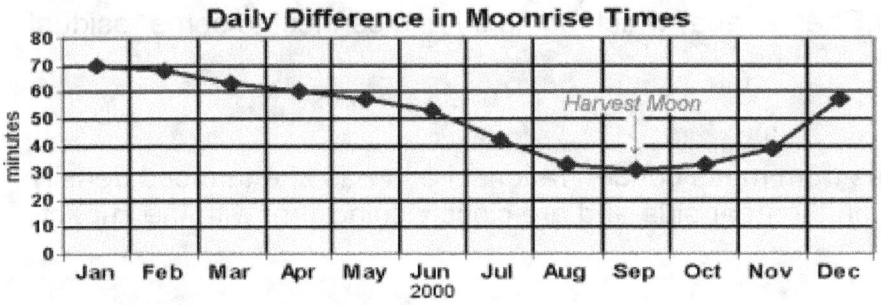

Figure 39 - Difference in Moonrise

Over time, we can see more than about 59% of the moon's face. It will 'nod' east-west slightly because of its elliptical orbit and its inclined orbit (inclined at 5° to the plane of the Earth's orbit) will produce a north-south 'movement'. There is also a diurnal difference. These motions are called *libration.*

---

[143] The Earth is more reflective - the Moon would be five times brighter if it had the same albedo as Earth.

The view of the Moon in a telescope could be reversed (in which case all objects will be reversed, of course).

I have already mentioned the Moon illusion, whereby the Moon on the horizon looks bigger than when it is higher up.

Lunar Eclipses

I have briefly mentioned these in Chapter 5 when I was discussing solar eclipses. During a lunar eclipse, the moon never becomes totally dark due to sunlight refracted through the earth's atmosphere (with its dust and cloud cover). The maximum length of an eclipse is 1 hour 44 mins.

Longitude Problem

The moon was also potentially useful for the solving of the 'Longitude Problem'[144], i.e. an accurate table of the Moon's positions allows ships to calculate their longitude.

This is often overlooked – because the solution was given definitively by Harrison and his timepiece(s)[145]. However these were expensive, so the study of 'lunars' continued for some years after Harrison had finally solved the problem.

Origin of the Moon

The favored theory nowadays is due to Hartmann and Davis. They envisaged a prototype earth being hit by an object of a size similar to Mars. The iron core of the impactor coalesced with that of the 'prototype' Earth and an enormous amount of material was ejected into orbit. This orbiting material coalesced into the Moon, a process which seems to have been assisted by the fact that the material was molten.

Now we have samples of Moon rock, we know its isotopic composition is similar to the Earth's, producing a good argument in favor of the so-called 'giant impact' theory. (An element is defined by the number of protons in its nucleus. The number of neutrons in the nucleus can vary, producing isotopes of the element. A sample of an element taken anywhere on Earth will have the same percentage of each constituent isotope. But that is a 'local' effect due the abundances in the solar material from which the Earth was formed and the fact that the material was thoroughly mixed during the Earth's molten stage. Other planets and bodies elsewhere in the Solar System can be expected to have different isotopic compositions for each element.)

Materials are commonly divided into volatiles (those that vaporize at low temperatures) and refractory materials (those that stay solid until high temperature). The Moon is depleted in volatiles as the theory requires. The

---

[144] Knowledge of the satellites of Jupiter could theoretically do the same job. Although naval tests were carried out, this procedure appears to be too impractical on a bobbing vessel. It was nevertheless used significantly for 'ordinary' mapping – you often hear how an early map using this technique 'reduced' the size of France considerably.
[145] which are on view at the Greenwich Observatory

Moon will be left with a small iron core but any magnetic field would soon have faded as the iron solidified.

The inclination of the Earth (at 23.5°) was possibly caused by this collision.

Advantages of Moon

The Earth has gained the following features from having a Moon.

- Its orbit has been stabilized. The inclination of other planets can alter markedly.

- It rotates more slowly – the day is longer

Space Probes

1959	
Luna 1[146]	Flew past the Moon
Luna 2	landed (or crashed to be more accurate) near to Archimedes
Luna 3	viewed far side
1964	
Ranger 7	crashed into Mare Nubium. First good close-range pictures of the moon
1966	
Luna 9	first soft landing, Oceanus Procellarum
Luna 10	first probe to go into lunar orbit, followed by other orbiters in 1966 and 1968
Surveyor 1	landed near Flamsteed Crater, Oceanus Procellarum, followed by 5 more landers in 1967 and 1968
Orbiter 1	mapped the Moon, followed by four more orbiters in 1966 and 1967
1968	
Zond 5	sent animals around Moon and returned to Earth
Zond 6	similar to Zond 5 but crashed on return to Earth
Apollo 8	Lunar orbital mission, Xmas 1968
1969	
Apollo 11	landed 20 July, stepped on to Moon 02:56 on 21 July (all times British time)
Luna 15	crashed on 21 July, attempt to return samples to Earth
Apollo 12	November. Oceanus Procellarum
1970	
Luna 16	landed in Sea of Fertility and returned sample to Earth
Luna 17	placed Lunokhod 1 on Mare Imbrium – this roved for 10.5km in 11 months
1971	
Apollo 14	January. Fra Mauro.
Apollo 15	July. Mare Imbrium : visited Hadley Rille, Apennines
1972	

---

[146] 'Luna' probes are sometimes referred to as 'Lunik'

Luna 20	Apollonius Highlands, south of Mare Crisium. Returned samples
Apollo 16	April. near Descartes
Apollo 17	December. Taurus-Luttrow, east of Sea of Serenity
1973	
Luna 21	placed Lunokhod 2 on LeMonnier – it traveled 37km in 5 months
1976	
Luna 24	Mare Crisium. collected and returned samples

Most Apollo missions landed near the equator and all took place during lunar morning when the Sun was low. At lunar noon the temperature would have been about 110°C.

The total of moon rocks returned to Earth amounted to 382 kg (USA) and 0.3kg (USSR).

Since Apollo there have obviously been no human landings but probes from several nations have been sent the Moon

Incidentally, the landing of early probes were verified by the Jodrell Bank radio telescope (who also 'stole' some early photos of the Moon). It seems that initially both the USA and the USSR has seriously been considering exploding nuclear bombs on the Moon to verify that they had reached there.

Chapter 8   Meteorites, Meteors, Comets, Variables, Miscellaneous

Meteorites

In general, most meteorites are composed of very similar material to asteroids. Both are composed of largely unaltered material from the solar nebula – asteroids never having reached a size where they can become molten and differentiate. Most meteorites are stony (96%) and could possibly actually originate from asteroid collisions. They are likely to be dated from about 4 billion years ago, or more.

Chondrites are a primitive type of meteorite, with carbonaceous chondrites being an even more primitive sub-type. These have been unmelted since formation although they almost all contain droplets of once-molten material – the chondrules from which they derive their name.

It's quite interesting that chondrites do also contain grains of material from other stars, minute diamonds and grains of silicon carbide typical of the stellar wind from red giants.

A small number of meteorites are iron (with about 5-15% by weight of nickel) formed by extreme melting processes in the body from which they originated. When etched with acid they display a *Widmanstätten Pattern*, which is not typical of terrestrial metal but is typical of nickel that has been cooled over an extremely long period (measured in millions of years)

Antarctica is a favorite site for gathering meteorites (where they are more likely to stand out). As of 1997, 19 of these meteorites were believed to have come from the Moon, and 12 from Mars

You might see the word 'meteoroid' used, which is the general name given to such types of objects when in space

Martian meteorites

A small number of meteorites are believed to come from Mars (and one hit the headlines in the tabloid press in 1996 because of claims that it contained evidence of fossilized bacteria[147]).

Most of them have ages of 0.2 to 1.3 billion years, instead of the usual 5 billion years old. And further they are definitely not 'ordinary' meteorites because they are composed, not of primitive unaltered material, but of basalt or similar volcanic rocks.

They are not from Earth because their isotopic composition is different. We have some idea of isotopic compositions from Martian landers, which are not incompatible with the isotopic compositions of these meteorites. For various

---

[147] The meteorite was called Allan Hills 84001 (ALH 84001). Allan Hills is an ice-field in Antarctica

reasons, other planets are excluded as their origin, leaving Mars as the favorite.

They are believed to have taken 16 million years to get here and hit Earth in the region of about 13,000 years ago

Meteors

Meteors are often classified as objects that don't reach the earth's surface whereas meteorites do. If I ignore micrometeorites, then I could state an alternative distinction, i.e. whereas meteorites stem from asteroids, meteors are produced by microscopic cometary dust hitting the atmosphere at enormous speeds and burning up fairly quickly. Typically they are moving at about 70 km/s and burn up above a height of 80 km.

Meteor showers are produced when the Earth actually enters the orbital path of a comet – an orbit strewn with dust. The best known shower is the Perseids every August, which is connected with the comet Swift-Tuttle.

The Perseids are a reliable shower (i.e. it occurs every year) and takes place when the skies can be clear in western Europe, and the weather warm. It has been noted in the historical records since at least 36 AD.

They receive their name because they appear to be coming from the constellation of Perseus, this being an optical effect. Observing in a 'serious way' by amateurs could involve several observers sitting in deckchairs placed in a circle, with one 'supervisor'. Sighted meteors are announced and it is up to the supervisor to decide whether it is a true Perseid (if so, it is logged) or a so-called 'sporadic' – a chance meteor coming from another direction in the sky.

When you think about the way the Earth is orbiting then you can visualize why the best time to observe is in the early hours – it is then that your location will be at the 'front' of the Earth's motion ploughing head long into the cometary dust.

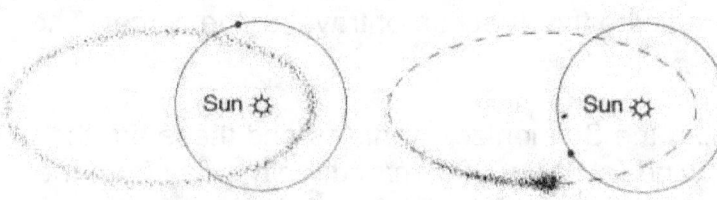

Figure 40 – Cometary Dust: uniform (l) and 'clumpy' (r)

Other showers tend to be less known because they are often clouded out, but some are also only occasional showers, e.g. the Leonids. In this latter case, the cometary dust is 'clumped'

around the orbit of the comet.

Meteor Shower	Date of Maximum	Comet
Perseids	August 12	1862III/Swift-Tuttle (seen again in 1992)
Seen from around 27/7 to 17/8. Consistent and best-known shower. Between 1864-66, Schiaparelli worked out the connection between the shower and Swift-Tuttle.		
Orionids	October 22	Halley (possibly)
Leonids	November 16	1866I Tempel-Tuttle
very inconsistent, with superb displays every 33 years or so, either side of the Comet's return. For example in 1966, with verified rates of 150,000 meteors per hour in places (for over 40 minutes). It was the Leonid shower of 1833 that led to the identification of specific meteor showers. Other good displays in 1799, 1866 and 1998 (with 1899 and 1933 being poor because the Earth missed the main swarm). 33 years is the orbital period of the comet itself.		
Geminids	December 13	

Comets

Comets are inherently small objects which nevertheless can engender great interest from the public when they are in the sky (e.g. Hale-Bopp, 1996[148]). Not all comets put on a good show, and periodic comets can vary on each of their appearances – for example, the most recent appearance of Halley's Comet in 1986 proved to be its least favorable on record[149]

The nucleus of a comet is only a few kilometers in diameter and is composed of ices, dust and small rocks and boulders. As it nears the Sun, evaporation[150] forms a coma. It is now realized that the comet is also immersed in a larger hydrogen cloud.

Particles in the coma are swept off to form two tails, both of which point away from the Sun – they do **not** indicate the direction of travel of the comet. They are

- *Ion tail* Ultra-violet from the Sun ionizes particles and these are then swept up in the solar wind (at about 300km/s) directly away from the Sun

---

[148] discovered in July 1995 by Alan Hale and Thomas Bopp
[149] and the next appearance in 2061 is predicted to be unfavorable
[150] whether evaporation is strictly speaking the correct word to use I am not sure – the ice goes direct from solid to gas, without a liquid phase

- *Dust tail* Dust is 'pushed' by radiation pressure from the Sun[151]. This tail will also point away from the Sun but not directly – it will be slightly curved

However, you will notice by looking at images of different comets that tails can vary.

The 'dirty snowball' model appeared in the 1950s and replaced the idea that they were formed of a loose aggregation of dust. Emissions do not stream off the entire nucleus but from those areas where the surface crust has been removed exposing the ice below.

A common test question asking for three differences between comets and other objects in the Solar System would expect the following three characteristics of comets to be mentioned

- they have highly elliptical orbits
- they orbit, in general, out of the plane of the Solar System
- they can also orbit the Sun in a retrograde fashion

For a long time comets were assumed to originate from the Oort Cloud, a collection of potential comets at a vast distance from the Sun (about a light year), where occasionally an odd one could be perturbed, e.g. by an object outside the Solar System, and thus be sent on those cometary paths that we can observe (there is not a shred of direct observational evidence for the cloud's existence). This theory has recently been modified such that the shorter-period comets (periods of about 200 years or less, although this limit is probably quite arbitrary) are assumed to originate from the recently-discovered Kuiper Belt – these comets might still have elliptical orbits but are less likely to fit the last two bulleted points just mentioned, in the sense that they will orbit closer to the plane of the Solar System and will tend to orbit in a prograde sense.

However, there is another factor to consider, namely the existence of the planets Jupiter and Saturn. These could maybe convert a long-period comet into a short-period one (we have an object only a few kilometers across pitted against the gravitational force of a giant planet). Conversely, some scenarios imagine the giant planets 'throwing' comets out of the early Solar System to actually form the Oort Cloud.

In 1995 a new method of designating comets came into being. In general. a long-period comet will be given a prefix of C and a short-period comet a prefix

---

[151] Not all dust will be swept up by the radiation. The larger particles will tend to remain in orbit and would form the material for meteor showers should the orbit intersect that of the Earth. Of those that become meteors, only the larger ones will burn up completely. The smaller ones will be able to get rid of the heat before suffering disintegration (smaller bodies have a larger ratio of surface area to volume and can thus lose heat more easily, cf small planets and satellites, and babies) and will thus survive to become micrometeorites. Considering all forms of dust, more material falls to Earth as dust rather than larger objects

of P. This prefix could be followed by the year of discovery, and further letters/numbers indicating in what part of the year it was discovered. For a 'P comet', the suffix will be preceded by a number (in a manner similar to the way asteroids are numbered). Thus you have the scientific designation - and a 'popular' name is an added extra. For example,

- Hale-Bopp is C/1995 O1 – a long period comet discovered in 1995. It was the first comet to be discovered within the half-month designated O (actually the second half of July) in the year 1995.

- Halley could be
    o 1P/Halley – a periodic comet given the number 1
    o 1P/1982 U1 (Halley) - if referring specifically to its most recent apparition (when it was the first comet discovered in the half-month referred to by the letter U (actually the second half of October))
    o 1P (Halley)
    o 1P.

The old systems might still be seen, e.g. 1973 XII or 1973f (Kohoutek) which would be rendered C/1973 E1 (Kohoutek) in the new system. Namings using added numerals, such as Tempel 2 (the second comet discovered by Temple), also belong to the old system.

The first person to study a comet in historical times appears to have been Tycho Brahe. He came to the conclusion that it was an extraterrestrial object – countering the prevailing idea that they were atmospheric effects. The 'Ptolemaists' required that they were not objects in space, otherwise they would crash through the individual transparent spheres on which the planets were located.

The first person to search for comets systematically was Messier who discovered thirteen[152]. He became so renowned for his work that he was awarded a pension, which unfortunately lapsed when his patron (de Saron of the Paris Parlament) was guillotined in 1794. Nevertheless there is a story that his patron was still willing to calculate the orbit of a new comet even while in prison.

In 1819, Encke studied the short-period comet which had first been seen in 1786. It has a period of 3.3 years but on each return it is always a couple of hours early. We now know that ejection of material during its orbit can act like a rocket and alter a comet's course slightly, or it can be perturbed by a planet

There were several 'Great Comets' in the nineteenth century – the one of 1811 is mentioned in some length in Tolstoy's 'War and Peace'. (Another interesting comet from this time was the short-period Biela's Comet which was observed to split into two, forming two separate comets on its next return, neither of which were seen again).

---

[152] this seems to be the most common number mentioned w.r.t Messier anyway

More recently there were a couple of 'great comets' in the 1990s, Hyukatake and Hale-Bopp.

A famous 'flop' was Kohoutek of 1973. In general, a comet is 'announced' when it is as far out as the asteroid belt – it is here when its ice would start to vaporize, producing the tell-tale signs of a comet. It appears that Kohoutek announced itself when material(s) other than ice, with a lower freezing point, were expelled leading its unwary observers to predict a 'great comet' as it approaches perihelion[153].

The most famous short period comet is Halley's Comet , with a period of about 76 years. It has its aphelion just the other side of Neptune (at 35AU), and its perihelion between Mercury and Venus This comet can now be tracked back via historical records to 240BC at least. It famously appears on the Bayeux Tapestry. Although it is a short-period comet, it orbits in a retrograde sense.

Isaac Newton was the first person to calculate the orbit of a comet, that of 1680. He then seems to have tired of that particular topic and handed his results over to Edmond Halley. Halley was able to extend Newton's work and produce data for about 20 or so comets. He noticed that 3 of these comets were very similar, those of 1682, 1607 and 1531, leading him to believe that they were one and the same object. He then predicted its return in 1752.

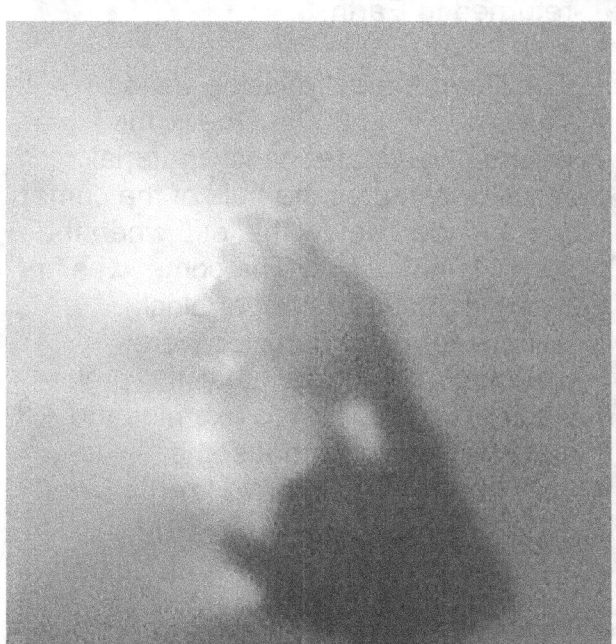

Figure 41 - The nucleus of Halley, as seen by Giotto

Nearer the time of its return, a group of French mathematicians recognized that Halley had made a couple of errors which cancelled each other out, producing essentially the correct answer. The period of Halley (and all periodic comets) is not constant, as we have already implied for Encke – the period of Halley varies between 68 and 80 years although in this case its 6-13 resonance with the planet Jupiter is all important for the variance in its period.

In 1986 it was visited by several probes, Giotto (Europe), Vega[154] (USSR) and Suisei and Sakigake (Japan). We now know its

---

[153] The minimum and maximum distances of an orbit around the Earth are referred to as the perigee and apogee. With respect to the Sun, the analogous terms are perihelion and aphelion.

[154] Giotto included a comet in one of his paintings (believed by some to be Halley). The word 'Vega' indicates that the craft first went to Venus (Ve) and then Halley (Ga) - Russian has no direct equivalent to the letter 'H'

nucleus is potato-shaped and measures 15 x 8 x 8 (km)[155], it is composed of 84% water. Only a small part of the nucleus would be 'active' at any one time, this region decaying by about 2cm per hour. Extrapolating to the entire comet, it loses about 2m of surface on every orbit – thus it is predicted to exist for another ¼ million years. The 1986 return was its 30th return since 240BC.

Apart from being fairly primitive by today's standards, the Halley probes also had the problem that Halley, by virtue of its retrograde orbit, was travelling in the opposite direction to the probes producing an enormous approach velocity. The first ever pictures of a cometary nucleus were taken, but only from about 600km and, all in all, showed very little detail.

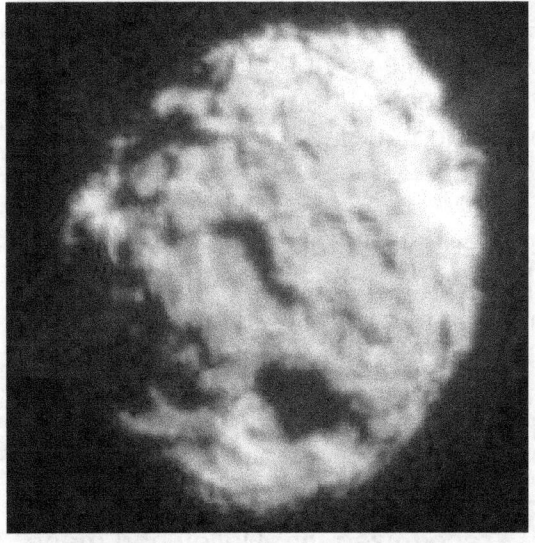

NASA's Deep Space 1 flew by Comet Borelly[156] in 2001.

The Starburst mission visited Wild-2 in 2004. The photos showed more detail and revealed features which looked like craters but were actually erosion areas. The mission received its name because it collected material from the comet. A 'sticky' panel gathered dust (in total about 2,000 particles) before its particular module detached and returned to Earth.

Figure 42 - the nucleus of Wild-2

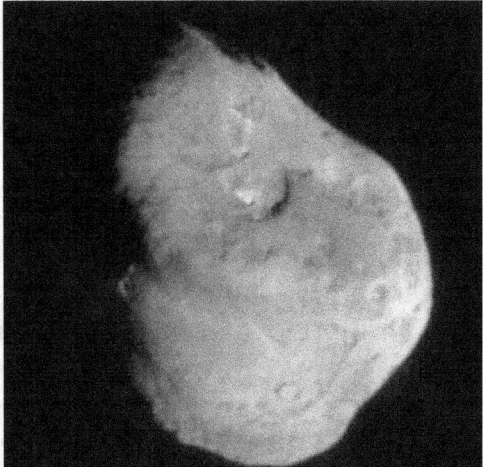

The Deep Impact mission visited Tempel 1[157], in 2005. One of the aims was to investigate the icy material which constitutes the bulk of the comet but is hidden below the 'dirt' when the comet is not active. The comet was hit by an impactor (a copper block). Various forms of particles were detected but unfortunately, the probe was moving too fast to examine the crater in the detail that some astronomers would have liked.

Figure 43 - nucleus of Tempel 1

---

[155] The nucleus of Hale-Bopp is assumed to be 10 times larger, meaning about 40km diameter. I say 'assumed' because just like all comets viewed from Earth the nucleus of Hale-Bopp was never viewed

[156] Comet Borrelly itself was discovered by Alphonse Borrelly on the evening of Dec. 28th,1904

[157] An original plan for a Halley mission involved a joint European/American mission – the European part going as far as Halley and the American part rendezvousing with Tempel 1. The Challenger crash ended those particular plans.

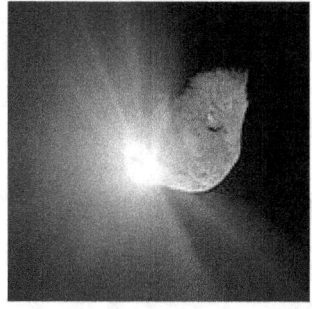

Tempel 1 was visited again in 2011 by the Next mission.

On the current photographs, some features that look like craters can be seen but these are former active regions which have eroded. The photos do show other features that also look like craters but are obviously not. What they are is not known at the moment.

Figure 44 - Tempel 1 after impact

The European probe, Rosetta, was launched in 2004 on its journey to Comet Churyumov-Gerasimenko and in 2014 it put a lander called Philae on to the surface, while the main probe orbited about 25km above the comet Unfortunately, the landing did not proceed as smoothly as planed and contact was lost with the lander after three days. It was originally planned to stay on the surface for over a year. witnessing the comet 'waking up' as it approached the Sun (its perihelion was reached in October 2015).

In 1994, comet Shoemaker-Levy 9[158] (which had been discovered in 1966) 'collided' with Jupiter. It was a strange object in the first place because it was orbiting around Jupiter, not the Sun. In 1993, it had already been noticed that the comet had fragmented (presumably because it went too close to Jupiter and got ripped apart by tidal forces) and at least 21 pieces hit Jupiter one at a time (at 60 km per second) over a six day period between 16 and 22 July 1994.

In 2007, Comet Holmes[159] brightened by 1 million times in a very short period, the largest known outburst by a comet. In only 42 hours in October 2007, the comet brightened from a magnitude of about 17 to about 2.8. There is no generally accepted reason for this behaviour (maybe a piece broke off).

Theoretically, the thrust of Jupiter could be enough to propel a comet out of the Solar System completely. In the same vein, I have heard of an estimate that 1 in every 3000 of our comets comes from another stellar system. So far we have seen 900 long-period comets (as well as 173 short-period ones). Theoretically a comet from outside our Solar System could be recognized by its orbit.

Variable Stars

This is a fruitful area for amateur astronomers. In my experience, talks by amateurs describing their own experiences are often well appreciated by the audience. Most variables are grouped under three classes – eclipsing, pulsating and eruptive. The nature of the last two classes can be closely related to the stars' position on the HR diagram.

---

[158] 9th short-period comet discovered by Carolyn and Gene Shoemaker and David Levy, although note that this sort of naming (i.e. numbering comets of the same name) is now out of favor
[159] **17P/Holmes** was discovered by the British amateur astronomer Edwin Holmes on November 6, 1892.

Strictly speaking, the 11-year solar cycle means the Sun's energy varies by about 0.1% over the cycle and would make the Sun a variable star but in reality we are thinking in terms of a larger variation.

Apart from novas, the first variable to be discovered was Mira[160], by Fabricius in 1596. Several months later it had disappeared but it was re-discovered in 1603.

Origin of Naming

Variable classes are typically given the name of the first star of that class to be discovered as a variable, e.g. T Tauri variables.

As regards a star itself, if a variable star already has a Greek letter name, the name will remain. Otherwise, the first variable found in a constellation would be given the letter R, the next S, and so on to the letter Z. Then the next star is named RR, then RS, and so on to RZ, SS to SZ, and so on to ZZ. Then the naming starts over at the beginning of the alphabet: AA, AB, and continuing on to QZ. This system (omitting the letter J) can accommodate 334 stars. The 335th variable would be called V335, followed by V336, etc. Some examples - SS Cygni, AZ Ursae Majoris, V338 Cephei.

Friedrich Argelander was the person responsible for this procedure. I have already explained the extended Bayer classification, which was only used as far as Q. Thus the reason for starting with a capital R.

There is an alternative designation, called the Harvard Designation (after Harvard College Observatory), which gives the variable's approximate coordinates for the year 1900. For example, the designation 0942+11 for R Leonis denotes an approximate position of right ascension of 09 hours 42 minutes and a declination of +11 degrees for the year 1900.

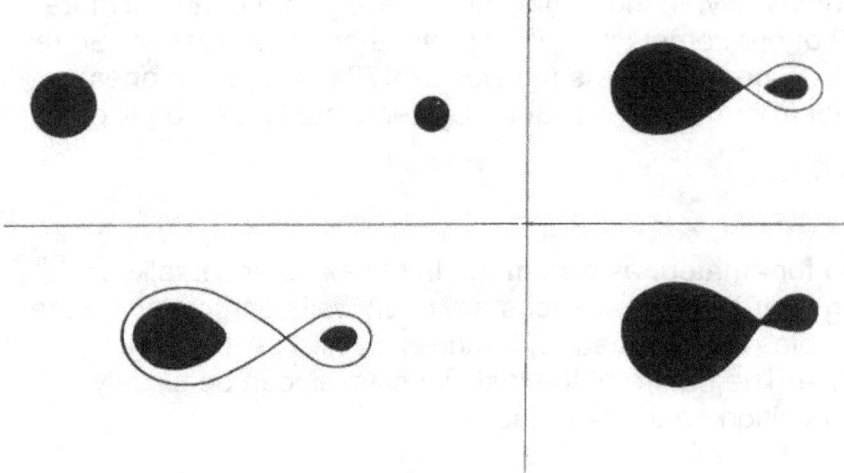

Figure 45 – Forms of binary systems

---

[160] Mira means 'the wonderful', a name seemingly stemming from it being the first variable to be seen

Top left in the above figure represents a 'normal' binary, while in the system at bottom left the stars are closer and more 'elongated' without any material being transferred between the two stars. The intersection of the 'figure of 8' is the 'First Lagrangian Point' (where the gravity of each body cancels each other out) while each separate 'lobe' delineates the star's 'Roche Limit' (within this limit, a body's own gravitation is capable of resisting the tidal effect of the other body). At top right, one star has filled its Roche limit – the system is 'semi-detached'. At bottom right we have a 'close binary' or 'contact binary'.

Eclipsing Variables

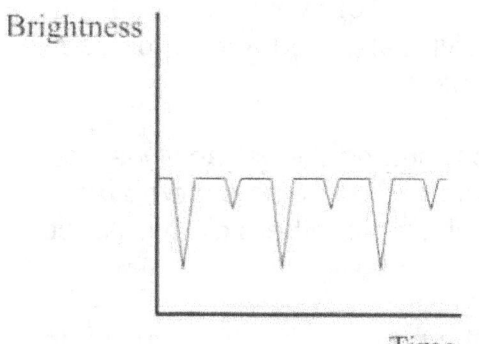

Algol (β-Perseii) is the best-known eclipsing binary, its brightness varying when one of the components goes behind the other. Its variability was discovered in 1669, but the explanation for it only came in 1782, given by John Goodricke (who also discovered the first Cepheid, δ-Cephei).

Figure 46 - Algol light curve

The variation in Algol itself is quite slight (between 2.1 and 3.4 every 3 days), but other Algol variables can vary more markedly.

Semi-detached binaries produce the β-Lyrae class, whose periods can stretch to 200 days but are most frequently about 24 hours.

Most contact binaries have components that are small – these produce the W Ursa Major class whose periods are always less than 24 hours and can even be as low as 1 hour. The components are usually similar to each other.

Pulsating Variables

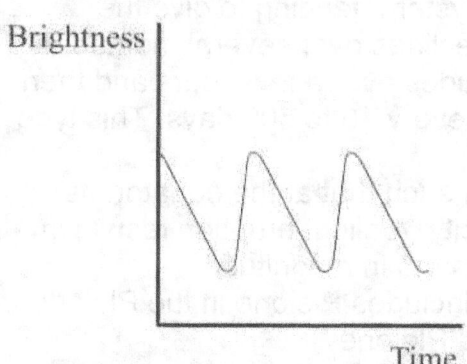

Cepheids are the best known pulsating stars. Their light curve is so regular that originally they were thought to be eclipsing variables. Their periods range from 2 to 50 days and, as discovered by Henrietta Leavitt, this period is linked with its intrinsic luminosity. Polaris was formerly classed as a Cepheid but recently the pulsations seem to have stopped. It was anyway a small-amplitude Cepheid, changing between 1.92 and 2.07 every 4 days. The prototype δ-Cephei varies between 3.48 and 4.37 over 5.4 days. – most well-known Cepheids vary by about one magnitude (over periods of about 1-60 days).

The 'classical' Cepheids just described are Population I stars. It was realised that Population II Cepheids behave differently. They are referred to as W Virginis types and are inherently two magnitudes less luminous than the classical Cepheids.

RR Lyrae although pulsating, these stars have no period-luminosity relationship. They are old population II stars and were first discovered in globular clusters – hence they were referred to as cluster variables. Their periods are always short – between 5 and 33 hours.

Mira or long-period variables the prototype is Mira or o (omicron) Ceti. They are cool red giants which increase by a magnitude of between 2.5 to 11 magnitudes over a period of 80-1000 days[161]. The light variation of Mira variables is not strictly periodic in that the individual period and amplitude of a Mira variable may vary from one cycle to the next.

They are Population I stars but the mechanism behind the variability is not well understood. Through mass loss they avoid exploding as supernovae. Some Miras are progenitors of planetary nebula, while others evolve directly to the white dwarf stage.

.
Because Mira variables are intrinsically bright, and because their amplitudes are large and their periods are long, they are particularly suited for visual observing. They are the most abundant class of variable.

Semi-regular variables these can resemble Mira stars but are less regular. They actually include several types. Examples include Betelgeuse, Antares and μ-Cepheii (the Garnet Star, named by William Herschel). The latter varies between 3.4 and 5.1 every couple of years and was the prototype of the semi-regular variables - it is a red supergiant

Eruptive variables

Nova this is generally accepted to be due to the nuclear detonation of accreted matter on a white dwarf in a binary system, tending to give the impression of a 'new star' in the sky. It then declines over several months. A Dwarf Nova is a type that rises by 4-5 magnitudes over a few hours and then declines, but re-occurs fairly regular between every 10 to 500 days. This type itself has sub-types like U Geminorum
γ-Cas type stars which are rotating very fast, such that at the equator, its velocity is only just less than the escape velocity. A slight eruption results in gas being expelled. Variation is often less than 0.5 in magnitude. γ-Cassiopeia is the brightest but the class also includes Pleione in the Pleiades. These ejections have produced a shell around Pleione.
UV Ceti, or flare stars are red dwarfs. A flare, similar in essence to a solar flare, will have a greater effect on the brightness of a red dwarf star than it does for the Sun

---

[161] Mira itself varies for the most part between about third to ninth magnitude over a period averaging 331 days. It is recorded to have reached first magnitude in 1779

Distance Scales

In general, different distances are measured using different techniques, forming a 'distance ladder', with the scaling of a particular method being based on another method used for lower distances. So an error lower down the ladder can introduce an error into the whole distance methodology.

The method at the 'bottom' of the ladder is parallax. This was used by Bessel and Herschel to make the first reliable estimates of distance. Hipparcos (High Precision Parallax Collecting Satellite), which was launched in 1989, has taken the limit of accuracy of this method out to 500 parsecs.

The Cepheid method is valid out almost as far as the Virgo Cluster

If I can just mention a few other methods, in order of distance (although there is generally a large overlap between these methods)

Tully-Fisher method : this relies on the assumption that there is a relationship between the intrinsic magnitude of a spiral galaxy and its rate of rotation.

Brightest Planetary Nebula : this works on the assumption that the brightest planetary nebula in each galaxy all have the same intrinsic magnitude

Type 1 supernovae : this method is based on the assumption that Type 1a supernovas are all of the same intrinsic luminosity at their peak

Brightest Cluster method : analogous to the planetary nebula method, this method assumes that for the richest galaxy clusters, the brightest galaxy in each cluster all have the same magnitude

Hubble's Law : as stated by the formula

$$z = H/c \; d$$

where z is the red shift, H is Hubble's Constant, c is the speed of light and d is the distance

The main problem with the last formula is that the value of Hubble' Constant is uncertain. Nevertheless, it can still be used to calculate ratios with more certainty, e.g. if the formula gives us a distance for object A that is twice the distance for object B, we know that A is twice as far away as B, irrespective of whether the calculated distances are correct or not.

Cosmic Rays

'Cosmic rays' are actually enormously fast, energetic particles. They move at a velocity comparable to the maximum achievable in Earth-based accelerators.

They were investigated from 1912 onwards by Victor Hess who detected them from balloons (an extremely hazardous procedure). They are mostly protons, although with many α-particles and about 3% electrons. Heavy elements are present, all the way up to Uranium. In general, cosmic rays hitting the atmosphere cause enormous showers of secondary particles which reach Earth. In general, only these secondaries reach ground level

Ionisation increases significantly above a height of about 1.5 km. In 1910, experiments were precise enough to realize that ionisation of the air at the top of the Eiffel Tower was significantly higher than the negligible amount expected if the X-rays[162] and γ-rays causing it originated from the ground.

Where cosmic rays come from exactly is uncertain. Being charged particles, their directions are altered by magnetic fields. Supernova and active galaxies are prime candidates, and some may come from outside the galaxy.

Incidentally, prior to the introduction of particle accelerators around 1953, particle physics relied on interactions caused by cosmic rays passing through cloud chambers by chance.

Sundials

A traditional sundial will have a 'gnomon', pointing at the pole star, and a horizontal dial – this horizontal dial will require that the spacing of hours are not equal. From what I have said previously, you can hopefully understand that the angle between the gnomon and the dial will be equal to the local latitude.

There are still problems

- It will show local time, not the recognized national time. For Britain, where Greenwich is in the East, western regions can be about 20 minutes behind Greenwich time
- Adjustments still need to be made for the Equation of Time (see below). In a nutshell, this adjustment is required because the Earth does not move along its orbit at a constant speed (see Kepler's Second Law).

Equation of Time

The Earth's orbit is an ellipse and the Earth moves at different speeds along this ellipse. There is an astronomical concept of a **mean Sun**, one that **does** move uniformly as though the Earth's orbit was circular and its speed constant. The real Sun will sometimes be in front of this 'mean Sun' and sometimes behind.

The time we use in everyday life is based on the behavior of this mean Sun, and variations between this 'mean Sun' time (the time we read off our clocks) and the time measured by the real Sun (e.g. on a sundial) can be as much as 16 minutes[163].

---

[162] X-rays had only been discovered in 1895.
[163] A sundial can be up to 16 minutes 'fast' and 14 minutes 'slow'

Stated as a formula (where the Equation of Time is the discrepancy between the two methods of calculating time)

Equation of Time = true solar time - mean solar time

Stated in another way

Equation of Time = Sundial Time - Clock Time

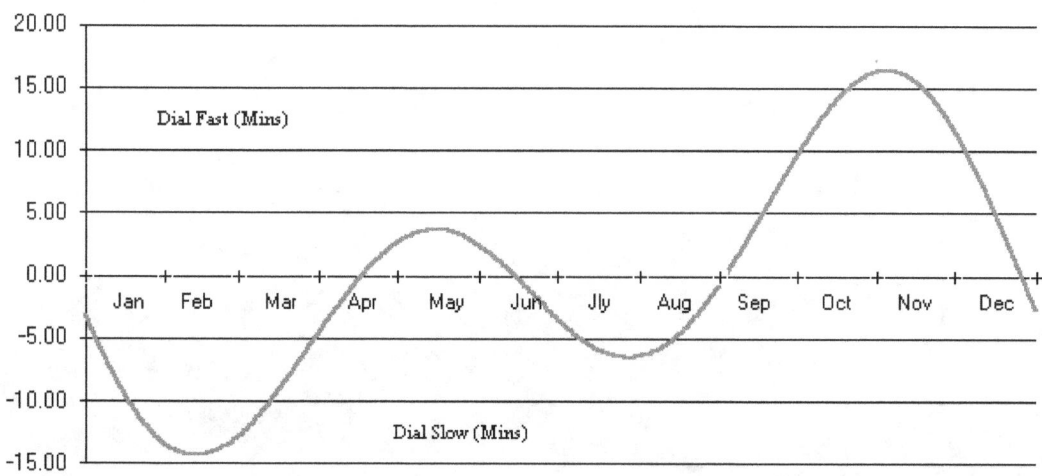

Analemma

If throughout the period of one year you were to plot the position of the Sun in the Sky at the same time of day, then you will produce a figure-of-eight figure - the Analemma.

Chapter 9   Cosmology

This chapter is the most theoretical (i.e. mathematical) but I'll do my best to explain the basics

Olber's Paradox

Although the appearance of the night-sky might appear quite logical, there is actually a paradox here – usually known as Olber's Paradox. The question is – why is the sky dark at night?[164]

In an infinite Universe[165], every direction would eventually meet a star – as a result, the sky should be blazing with light. If we were to visualize the stars as being arranged in 'spherical shells' around us, then the stars further away will indeed be weaker, but this will be compensated for by the fact that there will be a larger number of stars in each 'spherical shell' as the distance increases. Faintness is compensated by increased numbers. The total light will be the same from every 'spherical shell'.

I could state the paradox in a different way – why are there gaps of darkness between the stars? In fact, in view of what I said in the last paragraph, I could re-frame the question as : why is the sky so dark in the daytime?

Absorption by any intervening matter will not solve the problem because this matter will heat up due to this absorption and start emitting radiation on its own account.

The solution lies in the fact that whatever the nature of the Universe (it might be infinite), the actual *observable* Universe is *finite*.

This statement is a prediction from the so-called Big Bang Theory which states that the Universe was born something in the region of 15 billion years ago (the exact figure is subject to debate) and there has not been enough time since then for us to receive light from the entire Universe.

You don't need sophisticated equipment to view the evidence for a Big Bang – you see such evidence for yourself every time you look at the sky.

Basic Ideas about Curvature

As a result of the Theory of Relativity, we consider space and/or space-time to be curved. When we view things as being curved in 'real-life' we are looking at 2-dimensional surfaces and perceiving their curvature by virtue of looking at them in 3-dimensions. This need to view the curvature from 'one dimension higher' cannot obviously be used to 'view' the curvature of space (3

---

[164] This problem does not originate with Olbers. He wrote an essay on the subject but the name of the paradox was popularized during the 1950s by Hermann Bondi
[165] Some modern cosmological models do not describe an infinite Universe, but that does not detract from the argument being put forward here

dimensions) or space-time (4 dimensions). We have to resort to analogies to explain what is happening and to get the general ideas across– typically, explaining with a scenario in a lower dimension and asking you to extend the same ideas into a higher dimension.

An appropriate analogy is used in the influential book *Gravitation* by Misner, Thorpe and Wheeler. They have an ant crawling on the surface of an apple. If this ant has no concept of height[166], then it is constrained to move in two dimensions. How could it tell that the surface it is moving on is curved? It could start at a point and eventually find itself returning to the same point, although this is restricted a bit because not all 'models of curvature' have such a 'spherical' nature. Geometric methods would be more general

- initially-parallel lines will diverge or converge on curved surfaces

- angles of a triangle will not equal 180° on a curved surface. For example, you could draw a large triangle on the globe, which had a right angle at the North Pole and two right angles at the Equator

- the area of a circle will not equal πr^2 on a curved surface. If you can imagine a circle drawn with center at the North Pole of a paper globe – then cutting the circle out and trying to 'straighten' it out. You can only 'straighten it out' by tearing it – the area (represented by the paper) of the 'curved' circle is less than πr^2. An alternative form of curvature is commonly called 'saddle-shaped' – if you were to draw a circle on a saddle shape then you would find that the area of the circle is greater than πr^2.

A simple analogy in the same vein involves representing the Universe as an expanding balloon. The two-dimensional surface is representing the 3 dimensional space of the Universe. There is one **crucial** point to note here - we have to realize that the Universe is being modeled by the *surface* of the balloon. By this I mean the surface is everything – specifically

- there is no concept whatsoever of a traditional explosion with matter being blown into a pre-existing vacant space – it is space itself that is expanding (there is nothing else other than the surface of the balloon, representing the space of the Universe).

- concomitant with this, there is no center of the Big Bang any more than we can point to a particular position on the surface of the balloon and state that that is the center of the balloon's expansion.

With a bit of thought, you could hopefully accept the idea that information (i.e. light) from one part of the surface will be hindered from spreading across the entire surface by the expansion of the 'balloon' increasing the distance to be traveled.

---

[166] this is a thought experiment so we don't have to worry about the genuine capabilities of a real-life ant

If I can just append a few mathematical details, for completeness

- curvature has a more specific definition in mathematics than its 'everyday' definition. In mathematics the curved surface of a cylinder is actually flat – e.g. the label on a tin of food can be peeled off and laid flat without distortion. This is not so for the 'true' curvature that we are considering – I have already mentioned the idea of trying to flatten out a circle drawn on a sphere (and the way you can't do it perfectly)
- some portrayals of the balloon analogy have spots drawn on the balloon to represent galaxies. A proper analogy would need to have the galaxies represented by something like beads that will not themselves expand. If the galaxies themselves expanded then measuring equipment would also expand, and we would not be able to detect the expansion of the wider Universe.

Historical Aspects

The usual credit for discovering the expansion goes to Hubble who discovered that most of the 'recently-discovered' galaxies were red-shifted and therefore moving away from us[167]. He was able to produce a linear mathematical formula from his observations

$$V = Hd$$

where v is the radial velocity of a galaxy, H is Hubble's Constant[168] and d is the distance to the galaxy

.

He was anticipated by a couple of mathematicians, although their predictions seem to have been largely ignored (or not taken seriously) at the time. It was Einstein who noticed that his equations from General Relativity were predicting an expanding Universe – something he seems to have rejected on aesthetic grounds, leading him to modify his equations to produce a static Universe.

Alexander Friedmann put forward the idea of an expanding Universe in 1921. He published in Russian and died young, two features which presumably contributed to his ideas being overlooked. A Belgian priest, Lemaitre independently came to similar conclusions as Friedmann in 1927.

Friedmann Universes

---

[167] Vesto Slipher was the first person to identify galaxy redshifts and it a matter of conjecture why he did not come beat Hubble to the full conclusion (when Slipher started measuring the red-shifts, these galaxies were not even generally recognized as being separate galaxies external to our own).

[168] Hubble's Constant is the same everywhere but can change with time (in the most common model). The farther we look into space, the further back we are looking as well, so objects at a distance will be following a different value of Hubble's Constant than those closer (a situation which obviously complicates the calculation of the value of Hubble's Constant)

The theories of Friedmann Universes stem ultimately from Einstein's Equations as derived from his General Theory of Relativity. The equations are extremely intricate, but use is made of a standard technique in mathematics whereby simplifying assumptions are introduced, i.e. assumptions which simplify the equations involved. Any results derived are then obviously valid only under the introduced assumptions, and any resultant theory stands and falls on how valid these assumptions are.

The type of Universe described by Friedmann (and sometimes also co-attributed to Lemaitre[169]) rely on two assumptions

1. The Universe about us is spherically symmetric (or isotropic). This is a strong assumption in that it can be tested directly - is there a preferred direction in space or do all directions look the same?

2. The Universe is the same everywhere (or homogeneous). The upshot of this would be that the view from any other star would also show an expansion of the Universe identical to the one we see. In other words, the Universe will look isotropic from every other viewpoint within the Universe

These two assumptions form the Cosmological Principle

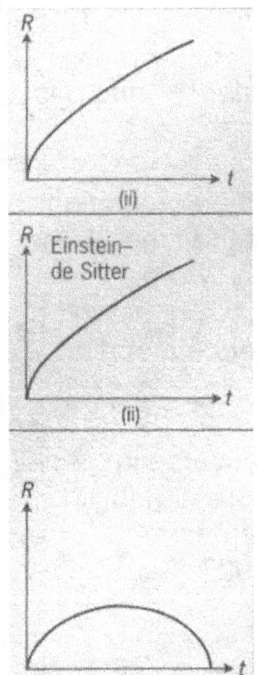

Figure 47 - Friedmann Universes

These assumptions simplify the (Einstein) equations, producing a relationship between the scale of the Universe and time. Depending on density of the Universe, there are two separate scenarios separated by a 'critical' scenario (as shown graphically in Figure 47). In all three of these scenarios, the Universe is slowing down (as you might expect because of the effect of gravity) . Uncertainties in actually measuring the density of the Universe produces debate as to which of these scenarios might be the 'right' one.

The middle diagram (the Einstein-de Sitter Universe) corresponds to the critical situation (when the Universe is said to have a 'critical' density).

The top diagram represents a **open** Universe where the density is less than the critical density – it slows down due the gravitational effect of the matter but nevertheless will expand for ever.

The bottom diagram represents a **closed** Universe where the density is higher than the critical density. The gravitational effect does not just slow the Universe but

---

[169] And often described as a Friedmann-Lemaitre-Robertson-Walker Universe, or any permutation of these names

actually reverses the direction of the expansion – producing a contraction into a 'Big Crunch'.

Although the Einstein-de Sitter Universe might be expected at first glance to purely represent the 'boundary' dividing the open and closed Universes and thus have little probability of actually representing the real Universe, particularly as estimates of mass from visible matter produces a density lower then the critical value (about 10% of that required for the Einstein-de Sitter Universe). Nevertheless, contemporary theories have tended to favor this option.

This is because the 'density deficit' only represents a very, very small deviation from the Einstein-De Sitter Universes at the time of the Big Bang itself when the scale was much 'smaller' A larger but still very small deviation at this early time would have produced a markedly larger gap between the current density of the Universe and the critical density. Rather than accept that the Universe was born with such a minute deviation from the Einstein-de Sitter, there has been a tendency to accept that the Universe is indeed Einstein-de Sitter.

This of course, in its turn, would mean that there is a lot of mass in the Universe that we haven't been able to detect visually – the so-called **dark matter or 'missing mass'**. Although we can't see it, we could theoretically detect it from its gravitational effect, although again uncertainties in measurement produce unclear results.

This has been the stimulus for much discussion over what type of objects matter could constitute this dark matter[170] - candidates stretch from conventional matter (dead stars etc.) through to exotic particles . The question of whether neutrinos have a mass or not is also a factor here.

Table 1 - Summary of Properties of Friedmann Universes

Open Universe	Einstein-de Sitter Universe	Closed Universe
density less than critical	density critical	density greater then critical
universe expands for ever	expansion tends to zero at infinity[171]	expansion slows to zero and reverses direction
infinite	infinite	finite
geometry is hyperbolic or 'saddle-shaped'	geometry is flat	geometry is spherical

Again, I'll just have to make a few comments on the above table

---

[170] By 'dark matter' we mean matter than we cannot see at all (but could detect gravitationally) – we are definitely not referring to dark areas of nebulas or anything of that nature
[171] this has been stated in 'abstract' and 'mathematical' language

- When the Universe is described as infinite, that is something that drops out of the mathematics and is not too 'understandable'. What we are saying is that the Universe is infinite right from the word 'go' and gets 'more infinite' as it expands. Whether you accept that or not, there is **definitely** no concept of a Universe becoming infinite through its permanent expansion. The closed universe is finite and spherical so that is the only one where you can talk about going right around the Universe and coming back to where you started from.

- all these three Universes are slowing down due to gravity, There is an analogy in considering an object that has been given an enormous (and single) impulse upwards from Earth. If the initial impulse is not large enough then the object will reverse direction and fall back to earth. A large enough impulse would send the object off, never to return, but nevertheless it would definitely slow down as it passed through the Earth's gravitational field[172].

However, the entire popular model just presented in this chapter has suffered a bit of a blow recently with evidence being advanced that the Universe is accelerating, not slowing down

Einstein's *gravitational constant*, which he had introduced to modify his equations (as just mentioned) was originally described (by Einstein himself) as a 'blunder' after Hubble's discoveries, but it seems to have come back in favor precisely because the equations can then produce accelerating Universes. However, claims from some quarters that Einstein was right all along seem to be exaggerated. If he turns out to be right, then he was right for the wrong reasons, which is effectively equivalent to being wrong (in my book).

Age of the Universe

You could theoretically use Hubble's Constant to obtain a crude estimate of the age of the Universe, as follows

$$T = \frac{1}{H_0}$$

where T = age of the universe and H_0 is the current value of Hubble's Constant (Hubble's Constant varies over time)

---

[172] This has been a thought experiment and, strictly speaking, only concerns a ballistic object i.e. something that has only been given a single 'thwack' on the Earth's surface, as opposed to a rocket that keeps its motors going for an extended period. However, Apollo rockets in orbit were only slightly further away from the Earth's center and the rocket burn that launched them from orbit towards the Moon was measured only in seconds, so it could be used as a good example of the general principles involved. They did slow down as they moved away from the earth (until they came under the influence of the Moon in a big way) – they would have slowed down whatever happened but if the burn had been too long they would have shot off, never to return.

This would be an over-estimate because Hubble's Constant was greater in the past. Debate over the value of Hubble's Constant extends into debate over the age of the Universe

Current estimates of Hubble's Constant are generally in the range 50-100 km/sec/Mpc[173]. A larger value of the constant reduces the age of the Universe, and vice-versa. A popularly quoted value of 55 for the value of Hubble's Constant would tend to produce an age for the Universes of about 18 billion years.

Sometimes there are 'problems' when estimates of the age of globular clusters turn out to be higher than the estimated age of the Universe itself.

Hot Big Bang

We have been talking about the Big Bang, but the idea that the Big Bang was actually hot requires some additional considerations.

In early times, radiation and matter were interacting with each other and only 'decoupled' when the temperature dropped to a certain level. From then on radiation was able to travel distances without being absorbed – in the same way that light from objects around us reach our eyes without being 'intercepted' and absorbed by matter on the way. The radiation from this decoupling period should exist as a 'Background Radiation' coming at us from all directions of the sky.

This background radiation would be black-body radiation, whose curve would thus be characteristic of a definite temperature. The original radiation will have been red-shifted due to the expansion of the Universe – although the red-shift is often interpreted as a Doppler shift, in reality the red-shift is caused by the wavelength of the radiation being stretched by the expansion of the Universe. By knowing the temperature of the Background Radiation now, we could then extrapolate back to when the decoupling took place and calculate its temperature at that epoch.

This Background Radiation was found in another classic case of Daedulus's Law[174]. It was discovered by Penzias and Wilson, employees of Bell Telephones, who were actually converting a telecommunications antenna for use in radio astronomy and were getting unexpected 'interference'.

The temperature of this Background Radiation was calculated to be 2.7K, which is obviously very low but extrapolation produced a temperature at decoupling of around 6000K, similar to the surface of the Sun.

---

[173] kilometers per second per megaparsec
[174] Daedulus's Law: all the great scientific discoveries are made by mistake when they were really looking for something else completely.

Note that although it is fashionable to state that the Background Radiation comes from the Big Bang, it actually stems from the decoupling event, 300 000 years after the Big Bang itself.

More recent research has looked for small fluctuations in the background radiation (COBE in 1992 and the more recent Wilkinson Microwave Anisotropy Probe (WMAP)). Although it is extremely 'smooth', there need to be small irregularities to allow condensation of galaxies.

The extreme conditions shortly after the Big Bang itself were ripe for element production but only existed for long enough to produce Hydrogen and Helium. Measurement of helium in the present Universes (which needs to be adjusted for helium production inside stars) therefore produces a constraint that cosmological models need to take account of.

Chapter 10   Telescopes and Observatories

Binoculars

Binoculars can be used for advanced astronomy. If I can just expand a bit on the remarks I made right at the very beginning, 'ordinary' binoculars are okay for beginners but astronomical binoculars are more efficient at light-gathering.

Telescopes

There are two main classifications of telescope – **refractors** and **reflectors**. I will just mention the basic features here without any absolute recommendations. Before buying a telescope you are always recommended to seek out the advice of experienced observers. The usual advice has always been to avoid making rash purchases of cheap telescopes which turn out to be useless for astronomical purposes.

The two primary qualities you are interested in are the

- **aperture size** which will mean the size of the objective lens (lens at the front) for a refractor, or mirror size for a reflector. This is a measure of its light gathering power

- **magnification**. The magnification is not the over-riding consideration – when you are magnifying an image you are spreading the same 'amount' of light over a larger area, making the image fainter, and if you don't have enough light to begin with, then the magnification is not going to produce as much detail as you might have expected.

Although reflectors superficially have a long history (its invention usually being attributed to Newton), it was only in 1856 that they began to be manufactured using silver-on-glass rather than with metal mirrors. Nevertheless, they were famously built and used earlier by William Herschel. Their most obvious advantages are that they can be built larger than the lens of a refractor (more light-gathering power) and are devoid of the color fringing that can affect an image produced by a refractor[175].

Traditionally, reflectors have come in two versions –

- *Newtonians* where the light is reflected from the objective on to a plane mirror which turns the light through a right angle, allowing viewing of the image from the side of the telescope.
- *Cassegrains*[176] where the light from the objective is 'bounced back' by a curved mirror through a hole in the objective, allowing viewing from

---

[175] although there are ways of getting around this in modern-day refractors
[176] Guillaume Cassegrain who invented the type in 1672

the back of the telescope (which is more useful for professional observatories for a start, rather than climbing on ladders to view the telescope from the side[177]). Another possible advantage is that they are shorter than Newtonians.

Nevertheless, Newtonians are popular with some observers. For one thing the only precise optical component needed is the curved mirror which some amateur constructors favor. And a portable 'Dobsonian[178]' version has become common by virtue of its modular form.

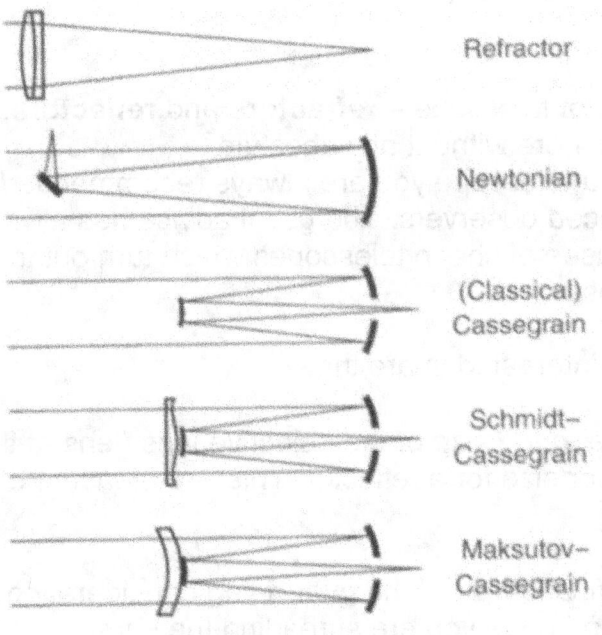

Figure 48 - Types of Common Telescope

Comparatively recently, Cassegrains have become popular in modified form. A Schmidt telescope (or camera, since this is what is was designed to be) has a 'corrector lens' fitted 'upstream' of the mirror, this mirror being spherical or near-spherical, in contrast to the parabolic shape of the traditional reflector mirror - a spherical shape is easier to produce than a parabolic shape. This instrument allows wide-angle fields to be viewed clearly. Applying these techniques to a Cassegrain produces the Schmidt-Cassegrain, as presented schematically at the left. This type was introduced for the first time in 1962.

Maksutov was a Soviet telescope-maker who used similar techniques to produce the Maksutov-Cassegrain. In this instrument all surfaces are spherical (or near-spherical – the exact details seem to be an industrial secret in some cases). The secondary mirror is just a metal spot on the corrector lens. This corrector is difficult to manufacture for larger telescopes, so this type is not used for professional purposes.

Traditionally, images in astronomical telescopes have been upside-down and back-to-front. Optically, it would be relatively simple to 'correct' the images but some light would be absorbed and lost in the correcting lenses. Nevertheless, nowadays it is quite common to have the image the right way up although still with left and right reversed.

---

[177] If a professional telescope is big enough, viewing can be carried out from a 'cage' at the prime focus, .i.e. inside the 'tube' roughly at the position that would be normally occupied by the secondary mirror

[178] After John Dobson of San Francisco, who popularized the type

Professional Observatories

The largest refractor is the Yerkes[179] Telescope of Chicago University with a lens of 102cm. diameter. This came into service in 1900 and its size (and weight of 250 kg) is generally considered to be the maximum possible for a refractor, anything larger tending to distort under its own weight.

The first modern reflector was Mount Wilson, California in 1917 (2.5m). In 1948 the 5m Hale telescope on Mount Palomar was inaugurated, 5 meters being generally considered to be the limit for a reflector because of the same problems of distortion due to weight

Although for many years the size of telescopes did not increase, their efficiency of detection did increase, most famously through the introduction of CCDs (charged-coupled devices) which can achieve efficiencies of up to 80% at some frequencies[180].

Nowadays, observatories are usually situated atop mountains in warmer (less cloudy) regions of the world. Tall mountains are above most clouds. Ideally they should also be surrounded by desert, meaning the air will be dry and thin

The Hubble Space Telescope gets above the atmosphere completely, which was once seen as the ultimate development but there has been a resurgence in Earth-based telescopes because of new computerized techniques, notably

- **active optics**. which compensates for distortion in the telescope itself
- **adaptive optics** which compensates for problems caused by the atmosphere

Active optics allows a computer to note distortions in the mirror and to restore the shape from behind as in, for example, the New Technology Telescope.

Another approach has been used by the Keck Telescope whereby a 10 meter mirror is actually composed of 36 hexagonal segments which are controlled individually by computer and can thus be combined into a perfect 10 meter objective

Adaptive optics allows a computer to compensate for twinkling due to the atmosphere. If there is an appropriate star close to the observed object, then since stars are always point sources, its twinkling can be observed and the image of the observed object adjusted on the assumption that it is being subjected to the same twinkling as the adjacent star. If no appropriate star is

---

[179] Although Yerkes is often mentioned in connection with his American interests, his companies were also heavily involved in the running of London Underground lines until they were nationalized in the 1930s

[180] changing from photography to CCDs is effectively the same as increasing the size of your telescope 10-fold

available, then an 'artificial star' can be produced by a laser, which activates sodium in the atmosphere[181] to produce the desired effect

The main professional sites include

- **Mauna Kea, Hawaii (4200m)** including *Keck* : 10m mirror composed of 36 hexagonal segments; *UKIRT (UK Infra Red Telescope Facility)* : 3.8m reflector; *James Clerk Maxwell Telescope* : submillimeter telescope, collaboration between Britain/Netherlands/Canada; *Canada-France-Hawaii Telescope* : optical/infra-red. Mauna Kea is well known as an exceptional site for infra-red observation

- **La Palma, Canary Islands (2400m)** including *William Herschel Telescope* : (4,2m), operated by Britain/Spain/Netherlands; *Isaac Newton Telescope* : (2.5m), *Jakobus Kapetyn Telescope*

- **La Silla, Chile (2400m)** (adjacent to the Atacama desert) : run by the European Southern Observatory

- **Cerro Paranal, Chile** run by the European Southern Observatory north of La Silla. Home of the Very Large Telescope Array, four 8m telescopes using active optics which can operate together to produce the equivalent of a single telescope of 16m. It can detect from near UV to mid IR. In 2004, it took the first photograph of a planet around another star.

- **Cerro Tololo (2200m)** : American observatory in Chile. 4m reflector

- **Las Campanas, Chile (2300m)** run by the Carnegie Institute of Washington

- **Siding Springs, New South Wales (1150m)** the Anglo-Australian Telescope (3.9m) has operated since 1973. The site is unusual for an astronomical location in that it rains (and trees grow). David Malin's photographs using this telescope are well-known. The site also houses the UK Schmidt, and others

- **Hubble Space Telescope** launched 1990 and 'repaired' in 1993.

Radio telescopes

Radio telescopes are the only other major type of telescope that can be used efficiently on the Earth's surface. The frequencies detected range from about 1mm up to 30cm, this upper limit being where radio waves start having trouble reaching Earth because of the ionosphere.

---

[181] at a height of about 90 km

Karl Jansky published a paper on the subject in 1932 but not too much notice seems to have been taken of it at the time. In 1937 Grote Reber of Illinois built a practical telescope but the technique appears to have received a major boost via the work done on radar during the war.

Modern radio telescopes tend to be arrays e.g. the Mullard Telescope based on a 5km stretch of the former railway line between Oxford and Cambridge and the Very Large Array in New Mexico laid out in the shape of the letter Y, each arm being 21 kilometers long[182]. In order to focus an image as efficiently as an optical device, a single, individual radio telescope would need to be 1 million times as large as an optical device[183].

Radio telescopes do not need to be dishes - cheaper devices can operate as radio telescopes, such as connected rod aerials[184] or wires next to each other

Jodrell Bank was opened in 1957 and although it is the best-known British telescope as far as the general public is concerned, there is also the previously mentioned Mullard Telescope of Cambridge University whose director, Martin Ryle, received the Nobel Prize for his work. The 100 meter Effelsberg Telescope in the Eifel Mountains is the largest steerable telescope but likewise seems not to be too well known to the (British) public at large

America has the NRAO (National Radio Astronomy Observatory) which operates telescopes at various sites, e.g. Charlottesville in Virginia, Greenbank and the Very Large Array.

A better-known example is the 305 meter installation at Arecibo, Puerto Rico. This makes use of a natural bowl, so the dish is fixed with the actual detector being strung on wires up above.

Since radio is lower energy radiation (less energetic than visual radiation), this radiation would be expected likewise to be emitted by lower energy sources. However, there was also radiation that could not so easily be identified with optical objects. Most of this turned out to be synchrotron radiation caused by high-speed electrons spiraling in a magnetic field at near-light speeds – enabling radio telescopes to probe highly-energetic objects.

A notable spectral line is the 21cm radiation produced by the electron in a hydrogen atom reversing its spin. Although this is a so-called 'forbidden' transition with a given hydrogen atom expected to emit this radiation only once every 12 million years on average, there is so much hydrogen in the Universe that this radiation is significant. Analysis of the Doppler Shift of this narrow line was able to show the rotation of spiral galaxies, by investigating the motion of its gas clouds

---

[182] actually one arm is only 18.9 km long; the array is made up of 27 telescopes.
[183] 'linking' radio telescopes at different locations together remotely has been a long-established technique (interferometry techniques)
[184] the telescope used by Antony Hewish and Jocelyn Bell to detect pulsars consisted of 2048 dipole antennas occupying about 2 hectares

A very early discovery was radio galaxies, which optical astronomers tried to identify with visual objects. In 1960, a radio galaxy was identified as the farthest object yet detected, spurring interest in a line of research which also led to the discovery of quasars.

Some of the best-known discrete[185] radio sources are

- *Cygnus A* : discovered in 1939 : the first radio galaxy to be discovered, possessing the largest apparent brightness of any source outside our galaxy .
- *Cassiopeia A* : supernova remnant of around 1660 which seems not to have been seen optically (or at least not recorded). It could have been dimmed by dark nebulosity prevalent in that area of the sky[186]
- *Taurus A* : Crab Nebula
- *Virgo A* : associated with a jet in M87, which is probably linked with a black hole
- *Centaurus A* : NGC 5128, the closest active galaxy
- *Sagittarius A* : at the center of the Milky Way (and a possible black hole)
- *Sagittarius B* : molecular cloud located close to the center of the Milky Way

As time progressed, the straightforward alphabetic designation of discrete sources was insufficient and many individual observatories starting drawing up their own catalogs using their own classification – nevertheless that of the Third Cambridge Catalog is one of the most commonly quoted, e.g. 3C 273 was the first quasar to be discovered.

Infra Red

At a basic level. infra-red can be detected by placing a thermometer in the 'invisible' region beyond the red end of a spectrum produced by a prism, or similar.

However, although this tends to connect infra-red in the public mind with heat, the term 'infra-red' actually extends across several different 'classifications' extending down to areas of very low temperatures (about 1 degree above absolute zero).. It is indeed still being transmitted by matter by virtue of its 'heat' but this 'heat' can be very, very low. Ironically, its detection on the Earth's surface is hindered by greenhouse gases

Emissions from molecules due to vibration and rotation between the constituent atoms are in the Infra-red range.

- **far red** is the current name given to frequencies close to the visual range. Unlike visual frequencies, it is not blocked by dust

---

[185] i.e. isolated distinct sources, as opposite to radiation emanating from a large area of the sky
[186] This supernova is not to be confused with Tycho Brahe's supernova of 1572, also in Cassiopeia

- **near infra-red** emitted by bodies of between 1000-2000 degrees, e.g. newly-born stars. There are some 'windows' in this range for telescopes situated on mountains, so it is possible to carry out IR research on the ground, e.g. UKIRT (United Kingdom Infra-Red Telescope) on Mauna Kea.

- **mid infra-red** can be observed using airborne observatories. The most famous has been the Kuiper Airborne Observatory, a Lockheed C-141, with 91 cm reflector[187] although this came to an end in 1995. The idea continues with SOFIA (Stratospheric Observatory for Infra-Red Astronomy), a converted Boeing 747. This radiation is emitted by objects with a temperature of about 100K, e.g. starburst galaxies, dust clouds around very young or very old stars. It is important to distinguish the 'real' signal from the 'glow' of infra-red from objects in the immediate surroundings, including the telescope itself. The telescope thus needs to be cooled, usually using low-temperature gas.

- **far infra-red** cannot be observed from the Earth's surface. It is emitted by some of the coldest objects in the universe, e.g. clouds of gas and dust. This includes dust surrounding high-energy regions (e.g. active galaxies), or around stars that have not yet initiated fusion (protostars). Far infra-red is a typical signature of dust – each grain is effectively emitting in this range as a black-body (if the temperature was higher the dust would be vaporized).

Infra-red can be used to probe regions which are visually obscured by dust. The fact that infra-red is scattered less by dust can be inferred to a certain extent by the way that red visual light is scattered less than blue light, producing red sunsets etc.

The first dedicated infra-red satellite was IRAS (Infra Red Astronomical Satellite), administered by the USA/Britain/Netherlands. It was launched in 1983 and operated for about a year, before the helium used to cool the shield to 16K, and the detectors themselves to 2K, had boiled off[188]. IRAS made many discoveries (about 350,000 sources). It hinted that some stars had planets and discovered starburst galaxies (about 75,000 of them). Many dust clouds were discovered in Orion.

Europe's ISO (Infrared Space Observatory) was launched in 1995. The Spitzer probe was launched in 2003 and orbits around the Sun. One of the advantages of a heliocentric orbit is that the mission can be expected to last longer and the probe is still operating in 2018. In 2005 it captured images of specific extra-solar planets, the existence of which had previously been inferred indirectly (and discoveries of this type have continued). Also in 2005, scientists announced that the Milky War has a more substantial bar structure

---

[187] It was the Kuiper Observatory that also discovered the rings of Uranus
[188] It is crucial in all infra-red detection to minimize the infra-red from the immediate surroundings, including the telescope itself

than previously thought   In 2006, it studied Cassiopeia A and found that the 'layer' structure of the star still existed in the expanding cloud. – lighter elements further out, being 'followed' by the heavier elements. This discovery stemmed from its ability to detect objects at lower temperatures.

Sub-Millimeter

The boundary between infra-red and radio is not clear-cut (which is also the case for other 'boundaries') and radiation in this general 'boundary zone' is often called sub-millimeter radiation.

From the ground, observation in this range is only possible 1) within a few windows and 2) from about four sites, of which Mauna Kea is the best known. The James Clerk Maxwell Telescope on Mauna Kea is the world's largest sub-millimeter telescope at 4.2m. It is owned by a consortium of Asian countries.

It detects molecules in cold dusty clouds – for molecules to exist in space they need to be shielded from ultra-violet. Since cold clouds are connected with star formation, these observations are therefore of relevance to this topic, especially since this stage in a star's life is currently poorly understood.

Ultra-Violet

The ozone layer blocks this radiation reaching earth, except for the longer wavelengths (which can cause sunburn in humans). So, for astronomers, the term 'ultra-violet' refer to those frequencies of 310 nanometers and less that are blocked by the ozone layer. And obviously satellites are needed to carry out research

Detectors were carried on very short rocket flights in the 60s (just as for X-rays and γ-rays). These were followed by a series of orbiting satellites. The International UV Explorer (IUE) was launched in 1978 by USA and ESA, and lasted until 1996 (during which time it was able to identify the progenitor of 1987A). The X-ray satellite Rosat has been able to do some work in UV (the frequency divisions can become a bit arbitrary here[189], as already hinted. The Extreme UV Explorer (EUVE) was launched in 1992, probing in short wavelength ultra-violet.

Topics of interest in UV include white dwarfs, cataclysmic variables and interstellar matter. It can detect molecules via absorption spectra of their electronic transitions, notably hydrogen molecules. CO can be detected more easily, but can also be used as a tracer for hydrogen molecules.

It also allows study of the hottest stars, of which there are some interesting examples in the constellation of Orion. **Rigel** is a blue supergiant with a surface temperature of 11 000K. However, **Saiph** is hotter with a surface temperature of 26 000K - much of its 'light' is actually being radiated as invisible ultraviolet light. **Bellatrix** is another blue-white supergiant at 21 500K. The belt stars of Alnitak, Alnilam and Mintaka all have similar temperatures.

---

[189] and at the other end some optical telescopes can carry out UV observations

- **Alnitak's** surface is at 31 000K. A stellar wind flows off at 2000 km/s colliding violently with other gases producing X-rays that can be detected from Earth. Surrounding Alnitak are several other clouds of interstellar gas, one of which contains the Horsehead Nebula.
- **Alnilam,** the brightest belt star (and the furthest), has a surface temperature of 25 000K. .
- **Mintaka**, the faintest of the three belt stars, has a surface temperature of 30 000K[190].

X-Rays

Even the most penetrating X-rays can only get down to 40 kms above ground level (gamma rays can get down to 10km – which is still above the peak of Everest). In fact, the vast bulk of X-radiation is absorbed within about 10 cm of 'normal' air.

Early experiments using captured V2 rockets resulted in the detection of X-rays from the Sun in 1948. These were the predicted X-rays from the hot corona

In 1962, a rocket was launched ostensibly to detect solar X-rays bounced off the Moon (which they failed to detect) but unexpectedly detected a powerful discrete source, Scorpius X-1. This was later found to be a neutron star surrounded by an accretion disk (recent research suggests that the accumulated matter comes from a binary component).

By 1966, 30 sources were known, including the Crab Nebula and M87

The first X-ray satellite was Uhuru, launched in 1970 off the coast of Kenya. Its discoveries marked the start of a revolution in research in this field.

So obviously we expect X-rays to emanate from very highly-energetic regions, e.g. potential black holes, active galactic nuclei, supernova remnants, binary stars containing a white dwarf, neutron star or black hole (stars at the end of their lives are an important area)..

It has also been found that all stars emit X-rays, presumably for the same reasons that our Sun emits X-rays from the high-energy corona.

- **Exosat** European, launched in 1983, lasted 3 years
- **Einstein Observatory**   launched in 1978, detected thousands of previously-unknown X-ray sources until 1981. It showed that all quasars emit X-rays and that all stars emit X-rays.
- **ROSAT** German/USA/British satellite of 1990 was engaged in the first complete survey of far ultra-violet and X-ray sources. It cataloged

---

[190] Mintaka is used as a "mount calibration" star by many amateur astronomers as it lies within a quarter of a degree of the celestial equator, which sees this star rising and setting almost exactly east and west

60,000 objects. It showed that the Pleiades contains 500 young and hot stars emitting in X-rays.
- **XMM-Newton** observatory European, launched 1999
- **Chandra** launched in 1999 to study X-rays and UV, and still operating.

Although the Trapezium in M45 is composed of four stars, X-rays show that one of them – theta -1C – is far the most powerful of them and is primarily responsible for lighting up the Orion Nebula.

Some discrete sources

- *Cygnus X-1* : 6-15 solar mass star in binary system. Candidate for the first black hole to be discovered in our galaxy.
- *Cygnus X-3* : binary supergiant and black hole, period 4.8 hours
- *Taurus X-1* : Crab Nebula
- *Hercules X*-1 : X-ray pulsar of period 1.24 seconds in a binary system of period 1.7 days
- *Centaurus X*-3 : binary system with X-ray source orbiting its companion every 2.09 days, the source itself varying every 4.8 seconds - an X-ray pulsar.

Gamma Rays

The first significant detection of gamma radiation was from the Sun – from solar flares. Gamma-ray emission from the plane of the Milky Way was first detected in 1967 by the OSO-3 satellite.

However, the field of gamma-ray astronomy took great leaps forward with the SAS-2[191] (NASA, 1972) and the COS-B (Europe, 1975-1982) satellites. They detected a band of radiation coming from the plane of the Milky Way, plus discrete sources associated with supernova remnants such as the Crab and Vela pulsars, as well as the 'original' quasar 3C 273. There is also a source at the center of the Milky Way, possibly related to a black hole. They confirmed the earlier findings of the gamma-ray background and produced the first detailed map of the sky at gamma-ray wavelengths.

Geminga (Gemini gamma-ray source) was one of these point sources whose nature was only identified in 1991, ironically with the help of X-ray emission detected by the ROSAT satellite (and previously by Einstein). Geminga is now generally considered to be the first example of a radio-quiet pulsar.

Gamma Ray Bursters are one of the great current mysteries – a short burst of gamma rays from an unknown (but necessarily very powerful) source. They were first made known in 1973 – they had been detected earlier by USAF satellites whose task was to monitor terrestrial nuclear bomb explosions. On average, one is detected every day. These bursts can last for a fraction of a second or for a few minutes. They do appear to come from far away in the

---

[191] Second Small Astronomy Satellite

Universe, and currently the most favored theory seems to be that at least some of them come from so-called *hypernova* explosions - supernovas creating black holes rather than neutron stars.

Kepler's Law

First Law *: Planets orbit on ellipses (with the sun at a focus of the ellipse)*

The distance of a planet from the Sun therefore varies. When the planet is furthest away it is at aphelion, and when it is closest it is at perihelion.

Second Law *: A radius vector connecting the Sun to a planet will sweep out equal areas in equal times*

The second Law is an early version of the Principle of Conservation of Angular Momentum. A classic example of this latter principle is an ice-skater who rotates faster when (s)he draws in initially-outstretched arms.

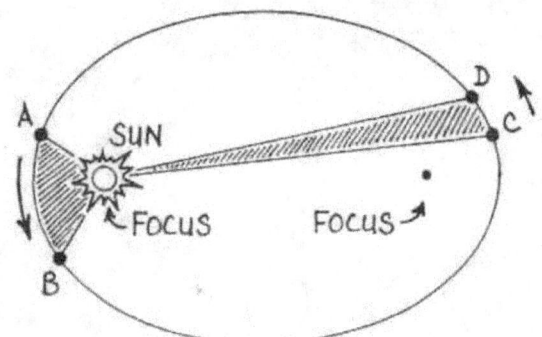

In the diagram, the two shaded regions have identical areas.
The period of time taken for the planet to travel between A and B is equivalent to the period of time it takes to travel between C and D.

Third Law *: The square of the orbital period is proportional to the cube of the semi-major axis*

Stated mathematically

$$T^2 = \alpha \, a^3$$

where T is the period, k is a constant and a is the semi-major axis

This can be converted to an equation thus

$$T^2 = k \, a^3$$

Where k is a constant

If the period is measured in years, and the semi-major axis in AU, then

$$T^2 = a^3$$

This third law can generalized to binary systems. Strictly speaking, a Sun/planet system is a binary system rotating about the common center of the system. The common center of the solar system is so close to the center of the Sun that we can approximate the situation by considering the Sun to be static (and also ignore the planet's mass on the assumption that it is negligible in comparison with the Sun's mass).